Alicja Ossowska

Highly Resolved Synthetic Aperture Radar with Beam Steering

Karlsruher Forschungsberichte
aus dem Institut für Hochfrequenztechnik und Elektronik

Herausgeber: Prof. Dr.-Ing. Thomas Zwick

Band 77

Highly Resolved Synthetic Aperture Radar with Beam Steering

by
Alicja Ossowska

KIT Scientific Publishing

Dissertation, Karlsruher Institut für Technologie (KIT)
Fakultät für Elektrotechnik und Informationstechnik, 2014

Impressum

KIT Scientific Publishing

Karlsruher Institut für Technologie (KIT)
KIT Scientific Publishing
Straße am Forum 2
D-76131 Karlsruhe

KIT Scientific Publishing is a registered trademark of Karlsruhe Institute of Technology. Reprint using the book cover is not allowed.

www.ksp.kit.edu

Print on Demand 2015

ISSN 1868-4696
ISBN 978-3-7315-0315-6
DOI 10.5445/KSP/1000044864

Vorwort des Herausgebers

Die Radarfernerkundung von Flugzeugen und Satelliten aus ist in den letzten Jahren zu einer der wichtigsten Informationsquellen über unsere Erde geworden. Die Charakterisierung der Vegetation, der Erdoberfläche bezüglich diverser Zustände wie Feuchte, die urbane Nutzung und die Auswirkungen von Erdbeben und Tsunamis können damit so präzise wie mit keinem anderen Verfahren wetterunabhängig erfasst werden. Die Entwicklung dieser Radarfernerkundungssysteme zielt auf immer höhere Auflösung und Vergrößerung der erfassten Gebiete. Zu diesen Zielen leistet die vorliegende Arbeit von Frau Ossowska einen wesentlichen Beitrag.

Die Arbeit analysiert die Designaspekte für mehrkanalige Synthetische Apertur Radargeräte mit digitaler Strahlformung (Digital Beamforming, DBF). Hierfür werden die nicht idealen Einflüsse der Systemkomponenten, wie die Nichtlinearitäten, anhand von Simulationen beschrieben. Ziel der Arbeit die Verifizierung Ihrer Ergebnisse mit einem sogenannten High Resolution Wide Swath SAR Demonstrator (HRWS) von EADS Astrium GmbH. Für die digitale Strahlformung zum Aufbau des HRWS SAR werden mehrere parallele Empfangskanäle benutzt. Zur Reduzierung des Prozessierungsaufwandes hat Frau Ossowska die Reduzierung der Quantisierung untersucht, insbesondere auch die 1Bit Quantisierung. Als interessantes Ergebnis zeigt sich, dass die 4Bit Quantisierung bezüglich Aufwand und Ergebnis optimal erscheint. Mit diesen Ergebnissen simuliert sie ein Flugzeug getragenes SAR und danach ein raumgestütztes SAR und untersucht die Möglichkeit für den Sliding Spotlight und den Hybrid Stripmap/Spotlight Mode. Die Ergebnisse werden in zukünftige Entwicklungen mit einfließen und die Fernerkundung weiter verbessern.

Ich wünsche Frau Ossowska auf Ihrem weiteren Weg, Erfolg in Forschung und Entwicklung und privat die Erfüllung ihrer Visionen.

Prof. Dr.-Ing. Dr. h.c. Dr.-ing. E.h. mult. Werner Wiesbeck

Forschungsberichte aus dem
Institut für Höchstfrequenztechnik und Elektronik (IHE)
der Universität Karlsruhe (TH) (ISSN 0942-2935)

Herausgeber: Prof. Dr.-Ing. Dr. h.c. Dr.-Ing. E.h. mult. Werner Wiesbeck

Band 1 Daniel Kähny
Modellierung und meßtechnische Verifikation polarimetrischer, mono- und bistatischer Radarsignaturen und deren Klassifizierung (1992)

Band 2 Eberhardt Heidrich
Theoretische und experimentelle Charakterisierung der polarimetrischen Strahlungs- und Streueigenschaften von Antennen (1992)

Band 3 Thomas Kürner
Charakterisierung digitaler Funksysteme mit einem breitbandigen Wellenausbreitungsmodell (1993)

Band 4 Jürgen Kehrbeck
Mikrowellen-Doppler-Sensor zur Geschwindigkeits- und Wegmessung - System-Modellierung und Verifikation (1993)

Band 5 Christian Bornkessel
Analyse und Optimierung der elektrodynamischen Eigenschaften von EMV-Absorberkammern durch numerische Feldberechnung (1994)

Band 6 Rainer Speck
Hochempfindliche Impedanzmessungen an Supraleiter / Festelektrolyt-Kontakten (1994)

Band 7 Edward Pillai
Derivation of Equivalent Circuits for Multilayer PCB and Chip Package Discontinuities Using Full Wave Models (1995)

Band 8 Dieter J. Cichon
Strahlenoptische Modellierung der Wellenausbreitung in urbanen Mikro- und Pikofunkzellen (1994)

Band 9 Gerd Gottwald
Numerische Analyse konformer Streifenleitungsantennen in mehrlagigen Zylindern mittels der Spektralbereichsmethode (1995)

Band 10 Norbert Geng
Modellierung der Ausbreitung elektromagnetischer Wellen in Funksystemen durch Lösung der parabolischen Approximation der Helmholtz-Gleichung (1996)

Band 11 Torsten C. Becker
Verfahren und Kriterien zur Planung von Gleichwellennetzen für den Digitalen Hörrundfunk DAB (Digital Audio Broadcasting) (1996)

Forschungsberichte aus dem
Institut für Höchstfrequenztechnik und Elektronik (IHE)
der Universität Karlsruhe (TH) (ISSN 0942-2935)

Band 12 Friedhelm Rostan
Dual polarisierte Microstrip-Patch-Arrays für zukünftige satellitengestützte SAR-Systeme (1996)

Band 13 Markus Demmler
Vektorkorrigiertes Großsignal-Meßsystem zur nichtlinearen Charakterisierung von Mikrowellentransistoren (1996)

Band 14 Andreas Froese
Elektrochemisches Phasengrenzverhalten von Supraleitern (1996)

Band 15 Jürgen v. Hagen
Wide Band Electromagnetic Aperture Coupling to a Cavity: An Integral Representation Based Model (1997)

Band 16 Ralf Pötzschke
Nanostrukturierung von Festkörperflächen durch elektrochemische Metallphasenbildung (1998)

Band 17 Jean Parlebas
Numerische Berechnung mehrlagiger dualer planarer Antennen mit koplanarer Speisung (1998)

Band 18 Frank Demmerle
Bikonische Antenne mit mehrmodiger Anregung für den räumlichen Mehrfachzugriff (SDMA) (1998)

Band 19 Eckard Steiger
Modellierung der Ausbreitung in extrakorporalen Therapien eingesetzter Ultraschallimpulse hoher Intensität (1998)

Band 20 Frederik Küchen
Auf Wellenausbreitungsmodellen basierende Planung terrestrischer COFDM-Gleichwellennetze für den mobilen Empfang (1998)

Band 21 Klaus Schmitt
Dreidimensionale, interferometrische Radarverfahren im Nahbereich und ihre meßtechnische Verifikation (1998)

Band 22 Frederik Küchen, Torsten C. Becker, Werner Wiesbeck
Grundlagen und Anwendungen von Planungswerkzeugen für den digitalen terrestrischen Rundfunk (1999)

Band 23 Thomas Zwick
Die Modellierung von richtungsaufgelösten Mehrwegegebäudefunkkanälen durch markierte Poisson-Prozesse (2000)

Forschungsberichte aus dem
Institut für Höchstfrequenztechnik und Elektronik (IHE)
der Universität Karlsruhe (TH) (ISSN 0942-2935)

Band 24 Dirk Didascalou
Ray-Optical Wave Propagation Modelling in Arbitrarily Shaped Tunnels (2000)

Band 25 Hans Rudolf
Increase of Information by Polarimetric Radar Systems (2000)

Band 26 Martin Döttling
Strahlenoptisches Wellenausbreitungsmodell und Systemstudien für den Satellitenmobilfunk (2000)

Band 27 Jens Haala
Analyse von Mikrowellenheizprozessen mittels selbstkonsistenter finiter Integrationsverfahren (2000)

Band 28 Eberhard Gschwendtner
Breitbandige Multifunktionsantennen für den konformen Einbau in Kraftfahrzeuge (2001)

Band 29 Dietmar Löffler
Breitbandige, zylinderkonforme Streifenleitungsantennen für den Einsatz in Kommunikation und Sensorik (2001)

Band 30 Xuemin Huang
Automatic Cell Planning for Mobile Network Design: Optimization Models and Algorithms (2001)

Band 31 Martin Fritzsche
Anwendung von Verfahren der Mustererkennung zur Detektion von Landminen mit Georadaren (2001)

Band 32 Siegfried Ginter
Selbstkonsistente Modellierung der Erhitzung von biologischem Gewebe durch hochintensiven Ultraschall (2002)

Band 33 Young Jin Park
Applications of Photonic Bandgap Structures with Arbitrary Surface Impedance to Luneburg Lenses for Automotive Radar (2002)

Band 34 Alexander Herschlein
Entwicklung numerischer Verfahren zur Feldberechnung konformer Antennen auf Oberflächen höherer Ordnung (2002)

Band 35 Ralph Schertlen
Mikrowellenprozessierung nanotechnologischer Strukturen am Beispiel von Zeolithen (2002)

Forschungsberichte aus dem
Institut für Höchstfrequenztechnik und Elektronik (IHE)
der Universität Karlsruhe (TH) (ISSN 0942-2935)

Band 36 Jürgen von Hagen
Numerical Algorithms for the Solution of Linear Systems of Equations Arising in Computational Electromagnetics (2002)

Band 37 Ying Zhang
Artificial Perfect Magnetic Conductor and its Application to Antennas (2003)

Band 38 Thomas M. Schäfer
Experimentelle und simulative Analyse der Funkwellenausbreitung in Kliniken (2003)

Band 39 Christian Fischer
Multistatisches Radar zur Lokalisierung von Objekten im Boden (2003)

Band 40 Yan C. Venot
Entwicklung und Integration eines Nahbereichsradarsensorsystems bei 76,5 GHz (2004)

Band 41 Christian Waldschmidt
Systemtheoretische und experimentelle Charakterisierung integrierbarer Antennenarrays (2004)

Band 42 Marwan Younis
Digital Beam-Forming for high Resolution Wide Swath Real and Synthetic Aperture Radar (2004)

Band 43 Jürgen Maurer
Strahlenoptisches Kanalmodell für die Fahrzeug-Fahrzeug-Funkkommunikation (2005)

Band 44 Florian Pivit
Multiband-Aperturantennen für Basisstationsanwendungen in rekonfigurierbaren Mobilfunksystemen (2005)

Band 45 Sergey Sevskiy
Multidirektionale logarithmisch-periodische Indoor-Basisstationsantennen (2006)

Band 46 Martin Fritz
Entwurf einer breitbandigen Leistungsendstufe für den Mobilfunk in Low Temperature Cofired Ceramic (2006)

Band 47 Christiane Kuhnert
Systemanalyse von Mehrantennen-Frontends (MIMO) (2006)

Band 48 Marco Liebler
Modellierung der dynamischen Wechselwirkungen von hoch-intensiven Ultraschallfeldern mit Kavitationsblasen (2006)

Forschungsberichte aus dem
Institut für Höchstfrequenztechnik und Elektronik (IHE)
der Universität Karlsruhe (TH) (ISSN 0942-2935)

Band 49 Thomas Dreyer
Systemmodellierung piezoelektrischer Sender zur Erzeugung hochintensiver Ultraschallimpulse für die medizinische Therapie (2006)

Band 50 Stephan Schulteis
Integration von Mehrantennensystemen in kleine mobile Geräte für multimediale Anwendungen (2007)

Band 51 Werner Sörgel
Charakterisierung von Antennen für die Ultra-Wideband-Technik (2007)

Band 52 Reiner Lenz
Hochpräzise, kalibrierte Transponder und Bodenempfänger für satellitengestützte SAR-Missionen (2007)

Band 53 Christoph Schwörer
Monolithisch integrierte HEMT-basierende Frequenzvervielfacher und Mischer oberhalb 100 GHz (2008)

Band 54 Karin Schuler
Intelligente Antennensysteme für Kraftfahrzeug-Nahbereichs-Radar-Sensorik (2007)

Band 55 Christian Römer
Slotted waveguide structures in phased array antennas (2008)

Fortführung als
"Karlsruher Forschungsberichte aus dem Institut für Hochfrequenztechnik und Elektronik" bei KIT Scientific Publishing
(ISSN 1868-4696)

Karlsruher Forschungsberichte aus dem
Institut für Hochfrequenztechnik und Elektronik
(ISSN 1868-4696)

Herausgeber: Prof. Dr.-Ing. Thomas Zwick

**Die Bände sind unter www.ksp.kit.edu als PDF frei verfügbar
oder als Druckausgabe bestellbar.**

Band 55 Sandra Knörzer
Funkkanalmodellierung für OFDM-Kommunikationssysteme bei Hochgeschwindigkeitszügen (2009)
ISBN 978-3-86644-361-7

Band 56 Thomas Fügen
Richtungsaufgelöste Kanalmodellierung und Systemstudien für Mehrantennensysteme in urbanen Gebieten (2009)
ISBN 978-3-86644-420-1

Band 57 Elena Pancera
Strategies for Time Domain Characterization of UWB Components and Systems (2009)
ISBN 978-3-86644-417-1

Band 58 Jens Timmermann
Systemanalyse und Optimierung der Ultrabreitband-Übertragung (2010)
ISBN 978-3-86644-460-7

Band 59 Juan Pontes
Analysis and Design of Multiple Element Antennas for Urban Communication (2010)
ISBN 978-3-86644-513-0

Band 60 Andreas Lambrecht
True-Time-Delay Beamforming für ultrabreitbandige Systeme hoher Leistung (2010)
ISBN 978-3-86644-522-2

Band 61 Grzegorz Adamiuk
Methoden zur Realisierung von dual-orthogonal, linear polarisierten Antennen für die UWB-Technik (2010)
ISBN 978-3-86644-573-4

Band 62 Jutta Kühn
AlGaN/GaN-HEMT Power Amplifiers with Optimized Power-Added Efficiency for X-Band Applications (2011)
ISBN 978-3-86644-615-1

Karlsruher Forschungsberichte aus dem Institut für Hochfrequenztechnik und Elektronik (ISSN 1868-4696)

Band 63 Małgorzata Janson
Hybride Funkkanalmodellierung für ultrabreitbandige MIMO-Systeme (2011)
ISBN 978-3-86644-639-7

Band 64 Mario Pauli
Dekontaminierung verseuchter Böden durch Mikrowellenheizung (2011)
ISBN 978-3-86644-696-0

Band 65 Thorsten Kayser
Feldtheoretische Modellierung der Materialprozessierung mit Mikrowellen im Durchlaufbetrieb (2011)
ISBN 978-3-86644-719-6

Band 66 Christian Andreas Sturm
Gemeinsame Realisierung von Radar-Sensorik und Funkkommunikation mit OFDM-Signalen (2012)
ISBN 978-3-86644-879-7

Band 67 Huaming Wu
Motion Compensation for Near-Range Synthetic Aperture Radar Applications (2012)
ISBN 978-3-86644-906-0

Band 68 Friederike Brendel
Millimeter-Wave Radio-over-Fiber Links based on Mode-Locked Laser Diodes (2013)
ISBN 978-3-86644-986-2

Band 69 Lars Reichardt
Methodik für den Entwurf von kapazitätsoptimierten Mehrantennensystemen am Fahrzeug (2013)
ISBN 978-3-7315-0047-6

Band 70 Stefan Beer
Methoden und Techniken zur Integration von 122 GHz Antennen in miniaturisierte Radarsensoren (2013)
ISBN 978-3-7315-0051-3

Band 71 Łukasz Zwirełło
Realization Limits of Impulse-Radio UWB Indoor Localization Systems (2013)
ISBN 978-3-7315-0114-5

Band 72 Xuyang Li
Body Matched Antennas for Microwave Medical Applications (2014)
ISBN 978-3-7315-0147-3

Karlsruher Forschungsberichte aus dem Institut für Hochfrequenztechnik und Elektronik (ISSN 1868-4696)

Band 73 Sebastian Diebold
Transistor- und Leitungsmodellierung zum Entwurf von monolithisch integrierten Leistungsverstärkern für den hohen Millimeterwellen-Frequenzbereich (2014)
ISBN 978-3-7315-0161-9

Band 74 Christian Rusch
Integrierte, planare Leckwellenantennen für 3D-Millimeterwellen-Radarsysteme basierend auf dem holografischen Prinzip (2014)
ISBN 978-3-7315-0234-0

Band 75 Marlene Harter
Dreidimensional bildgebendes Radarsystem mit digitaler Strahlformung für industrielle Anwendungen (2014)
ISBN 978-3-7315-0249-4

Band 76 Michael A. Baldauf
Abhängigkeit der Exposition von der Zellgröße beim Mobilfunk unter Gewährleistung der Versorgung (2015)
ISBN 978-3-7315-0308-8

Band 77 Alicja Ossowska
Highly Resolved Synthetic Aperture Radar with Beam Steering (2015)
ISBN 978-3-7315-0315-6

Highly Resolved Synthetic Aperture Radar with Beam Steering

Zur Erlangung des akademischen Grades eines

DOKTOR-INGENIEURS

von der Fakultät für
Elektrotechnik und Informationstechnik
des Karlsruher Instituts für Technologie (KIT)

genehmigte

DISSERTATION

von

Dipl.-Ing. Alicja Barbara Ossowska
geb. in Gdańsk

Tag der mündlichen Prüfung: 27. 02. 2014
Hauptreferent: Prof. Dr.-Ing. Dr. h.c. Dr.-Ing. E.h. mult. Werner Wiesbeck
Korreferent: Prof. Dr. rer. nat. Helmut Süß

Vorwort

An erster Stelle möchte ich mich bei Herrn Prof. Dr.-Ing. Dr. h.c. Dr.-Ing. E.h. mult. Werner Wiesbeck für die wertvolle und hilfreiche Unterstützung meiner Forschungsarbeit und die Übernahme des Hauptreferats bedanken. Ebenso danke ich Herrn Prof. Dr. rer. nat. Helmut Süß für die Übernahme des Korreferats. Weiter mein besonderer Dank an Herrn Prof. Dr.-Ing. Thomas Zwick für die hilfreichen Diskussionen und seine Motivierung.

Weiter möchte ich mich bei den Kollegen und Mitarbeitern des Instituts für Hochfrequenztechnik und Elektronik der Universität Karlsruhe und der Abteilung Aufklärung und Sicherheit des Instituts für Hochfrequenztechnik und Radarsysteme des Deutschen Zentrum für Luft- und Raumfahrt e.V., und bei allen meinen Studenten, die alle wesentlich zum Gelingen dieser Arbeit beigetragen haben herzlich bedanken. Ganz besonders möchte ich mich bei Herrn Dr.-Ing. Christian Sturm und Dr.-Ing. Marwan Younis für die vielen hilfreichen Anregungen und Diskussionen, die Motivation und die konstruktive und sorgfältige Durchsicht des Manuskriptes bedanken. Ich möchte mich auch bei Herrn Dipl.-Ing. Tom Schipper für die vielen Diskussionen und die Hilfsbereitschaft herzlich bedanken. Weiter mein besonderer Dank an Elvira Munz für die Hilfe bei der Korrektur.

Darüber hinaus bedanke ich mich bei Dr.-Ing. Udo Habberland, Dr.-Ing. Urs Lübbert und Dr.-Ing. Bernhard Schürmann für Unterstützung.

Ebenso herzlich bedanke ich mich bei meinen Eltern und besonders bei meinen Mann, Paul Ortyl, für die stetige Unterstützung und Motivation während der Promotion.

Karlsruhe, Dezember 2014
Alicja Ossowska

Zusammenfassung

Diese Arbeit beschäftigt sich mit einem hoch auflösenden Radar mit synthetischer Apertur (Synthetic Aperture Radar - SAR). Der allgemeine Trend in der Radarentwicklung geht in Richtung hoher Auflösungen und großer Abbildungsflächen. Eine Vielzahl neuer Applikationen erfordert flexiblere Radarsysteme, und es werden unterschiedliche Anforderungen an zivile und militärische Systeme gestellt. Der Schwerpunkt des zivilen SAR sind große Abbildungsflächen mit mittleren Auflösungen. Dasselbe Gebiet muss zur Beurteilung von Änderungen mehrfach zu unterschiedlichen Zeiten abgebildet werden. Militärische Anwendungen hingegen erfordern in der Regel Abbildungen von relativ kleinen Flächen mit sehr hoher Auflösung. Das Radarbild muss in sehr kurzer Zeit aufgenommen und verarbeitet werden. SAR-Systeme haben zwei Hauptbestandteile: die Hardware und der SAR-Prozessor. Jeder dieser Bestandteile ist von großer Bedeutung für die Leistung des Radars. Der erste Teil dieser Arbeit beschäftigt sich mit den möglichen Auswirkungen verschiedener Effekte in der Hardware des SAR-Systems. Der zweite Teil der Arbeit beschreibt SAR-Prozessoren.

Das High-Resolution Wide-Swath SAR (HRWS SAR) nutzt mehrere Empfangskanäle, um den Gegensatz zwischen Azimutauflösung und Streifenbreite zu überwinden. In dieser Arbeit wird der Einfluss des Empfangszweigs auf das Radarbild untersucht, und es wird eine neue Darstellung der Elemente des Empfängers und der möglichen Funktionen vorgeschlagen. Mit dieser Darstellung können alle Elemente und Funktionen im Empfänger einheitlich und einfach beschrieben werden; zudem ist es möglich, sowohl die Inphase- als auch die Quadraturkomponente des Empfangssignals gleichzeitig mit einem Modell zu zeigen. Dies ist von Vorteil für Simulationen. Es wird eine Beispielsimulation eines SAR-Systems mit Amplituden- und Phasenkopplung mit diesem Darstellungsmodell vorgestellt.

Leistungsverluste können zudem durch falsch positionierte Elemente der Gruppenantennen verursacht werden. Es können geringfügige Verschiebungen der Position von Arrayelementen bedingt durch mechanischen Stress oder Produktionstoleranzen vorkommen. In dieser Arbeit werden zwei mögliche

Verzerrungen der Array-Oberfläche betrachtet. Eine symmetrische Verzerrung, bei der die Ränder in der gleichen Richtung verschoben sind, sowie eine unsymmetrische Verzerrung, bei der die Ränder in gegenläufiger Richtung verschoben sind. Die Simulation ergibt, dass die symmetrische Verzerrung eine größere Auswirkung auf das Antennenpattern hat, während die unsymmetrische Verzerrung zu einer Hauptkeulenverschiebung führt. Dies ist für ein SAR-System besonders störend, wie die Ergebnisse der SAR-Simulation auch bestätigen.

Jede einzelne Empfangsantenne verfügt über einen eigenen Empfänger. Zusammen mit der hohen Auflösung ergeben sich hierdurch sehr große Mengen an Empfangsdaten, die es zu speichern und zu verarbeiten gilt. Eine Möglichkeit, die Datenmenge zu reduzieren, ist die Quantisierung mit wenigen Bits. Von besonderem Interesse ist das Single bit-Quantisierungsprinzip. Die HRWS SAR-Simulation zeigt die Ergebnisse der Systemleistung für das Ausgangsignal-zu-Rausch-Verhältnis (SNR) abhängig vom Eingangssignal-zu-Rausch-Verhältnis, der Anzahl der Empfangskanäle und der jeweiligen Anzahl an Quantisierungsbits. Besonderes Augenmerk wurde auf die Single bit-Quantisierung gelegt. Wichtig für die Rekonstruktion des Signals bei der Single bit-Quantisierung ist das additive weiße Rauschen. Obwohl es möglich ist, ein Single bit quantisiertes Signal durch digitales Beam Forming zu rekonstruieren, muss man mit einem Systemverlust von ca. 2 dB SNR rechnen. Die optimale Anzahl von Quantisierungsbits ist vier. Bei dieser Anzahl ist der Quantisierungsverlust vernachlässigbar.

Der zweite Teil der Arbeit betrifft die Verarbeitung eines hochauflösenden SAR-Systems. Die unverarbeiteten Rohdaten ähneln weißem Rauschen und eine Zielerkennung anhand von Rohdaten ist nicht möglich. Dazu müssen die Rohdaten erst verarbeitet werden. Die bekanntesten SAR-Verarbeitungsalgorithmen sind der Chirp Scaling- und der Omega-K-Algorithmus. In dieser Arbeit wird ein besonderer Arbeitsmodus beschrieben: der Sliding Spotlight Mode. Der Strip-map Mode ist ein verbreiteter Modus, der eine unbegrenzte Länge der Bildstreifen erlaubt. Die Azimutauflösung ist in diesem Modus jedoch auf die Hälfte der effektiven Antennenlänge begrenzt. Dies kann man mit dem Spotlight Mode verbessern, bei dem die Antennenkeule so geschwenkt wird, dass sie über den gesamten Datenerfassungszeitraum denselben Bereich beleuchtet. Allerdings hat dies den Nachteil, dass die Streifenlänge sehr begrenzt ist. Der Sliding Spotlight Mode ist ein Kompromiss zwischen der hohen Azimutauflösung der Spotlight Mode und der Länge

des beleuchteten Streifens in der Strip-map Mode. In dieser Arbeit werden zwei neue Algorithmen für den Sliding Spotlight Mode vorgestellt und verifiziert. Der erste Prozessor basiert auf dem Chirp Scaling-Algorithmus, der Zweite auf dem Omega-K-Algorithmus. Beide Algorithmen werden für Airborne-SAR-Systeme beschrieben und dann für Spaceborne-SAR-Systeme entsprechend modifiziert. Der vorgestellte Sliding Spotlight Chirp Scaling basierte Algorithmus ist geeignet für Azimutauflösungen bis ca. $0,5$ Meter. Der Omega-K-Algorithmus für Sliding Spotlight-Modus ist sogar für Azimutauflösungen bis ca. $0,3$ Meter geeignet (belegt durch Simulation).
In dieser Arbeit werden verschiedene Aspekte eines SAR-Systems mit Antennenkeulenschwenkung vorgestellt. Es wird auf zwei unterschiedliche Möglichkeiten der Schwenkung der Antennenkeule zur Verbesserung der Leistung des Systems eingegangen: Das HRWS SAR und der Sliding Spotlight Modus. Das HRWS SAR nutzt digitales Beam Forming, um die Gegensätze der Azimutauflösung und der Streifenbreite zu überwinden. Der Sliding Spotlight Mode ermöglicht einen Kompromiss zwischen der Azimutauflösung und der Streifenlänge des Strip-map-Modus. Es wird gezeigt, dass sowohl die Hardware als auch die Signalverarbeitung großen Einfluss auf die Leistung des SAR-Systems haben.

Summary

The present work deals with a highly resolved radar with a synthetic aperture (synthetic aperture radar - SAR), which uses a beam steering to improve performance. The general trend in radar development is higher resolutions and wider illumination areas. New applications require the radar system to be more flexible to cope with various demands. There are different requirements for civil and military systems. The main focus in a civil SAR is on a large illumination area with a moderate resolution. The area is to be revisited several times to assess changes. Military applications usually require a very high resolution image of relatively small areas. Radar images must be acquired and processed in a very short time. Radar systems consist of two integral parts: the hardware and the SAR processor. Either part is important to achieve high performance. The first part of this work deals with the influence of various effects occurring in the hardware of the SAR system. The second part of this work describes SAR processors.

The High-Resolution Wide-Swath SAR (HRWS SAR) consists of multiple receive channels. This configuration helps to resolve the contradiction between the azimuth resolution and the swath width. The influence of the receive path on the radar image is analyzed in this work. A new representation for elements of the receive front-end is proposed. The representation can describe all elements and effects of the receiver or transmitter in a simple and consistent way. Both the inphase and quadrature components of the receive signal can be described at the same time. An example simulation is presented using the proposed description model of a SAR system with an amplitude and phase imbalance.

Another possible source of performance loss is a misplacement of antenna array elements. Minor position errors of sub-antennas in the array can be caused by mechanical stress or production errors. Two possible position error types are considered in this work. In one type (in this work referred to as symmetrical position error) the edges of the array bend in the same direction. The other type called unsymmetrical position error results in edges bending in opposite direction. The simulation of a single receive panel shows the symmetrical position error has a stronger impact on the antenna array beam pattern. The unsymmetrical error induces a shift in the direction of the main lobe of the antenna characteristic. This is particularly disturbing for SAR systems. The simulation results presented confirm this.

Each receive antenna has its own receive front-end. Consequently, also considering the high resolution, a very large amount of receive data is obtained that needs to be saved and processed. One possibility to reduce the amount of data is using less quantization bits. Especially interesting is the single bit quantization case. The HRWS SAR simulation shows system performance results for the output signal-to-noise ratio against the input signal-to-noise ratio, the number of receive channels and the varying number of quantization bits. A special focus was set to single bit quantization. When reconstructing the single bit quantized signal, additive white noise on the input of the system is important. Although it is possible to reconstruct the single bit quantized signal by means of beam forming, a 2dB loss of the system performance must be accounted for. The optimal number of quantization bits is four. The performance loss due to quantization is negligible in this case.

The second part of the present work deals with the processing of a highly resolved SAR system. The unprocessed raw radar data is quite similar to white noise. It is not possible to recognize any targets. The raw data must be processed first. The best known SAR processing algorithms are the chirp scaling and the omega-K algorithms. In this work, a special working mode called sliding spotlight mode is considered. The strip-map mode is a very common mode allowing for continuous illumination of the observed area. The azimuth resolution in this mode is limited to half the effective antenna length. This can be improved using the spotlight mode, where the antenna is steered to illuminate the same spot over the whole data acquisition time. The disadvantage of the spotlight mode is the limited length of the illuminated area. The sliding spotlight mode is a compromise between the high resolution of the spotlight mode and the great illumination length of the strip-map mode. In this work, two processing algorithms for sliding spotlight mode are proposed and verified. The first processor is based on the chirp scaling algorithm, the second on the omega-K algorithm. Both algorithms are described for an airborne SAR, and then modified accordingly to cope with a spaceborne SAR system. The proposed chirp scaling based algorithm provides a good performance up to resolutions of about 0.5 meters. The omega-K algorithm for sliding spotlight mode provides a good performance for radar data with resolutions up to

0.3 meters, which is proved by a simulation.
Various aspects of a highly resolved SAR system are presented in this work. Two systems: HRWS SAR and sliding spotlight mode, which use beam steering to improve system performance, are considered. The HRWS SAR resolves the contradiction between the azimuth resolution and the swath width by beam forming. The sliding spotlight mode is a compromise between the high resolution of the spotlight and the swath length of the strip-map mode. It is shown that the performance of the SAR system is strongly influenced by both the hardware and the signal processing.

Contents

Abbreviations and Symbols

List of Constants

c_0	speed of light in vacuum 2.997926 m/s
k	Boltzmann's constant 1.38 J/K
R_{Earth}	mean Earth radius 6378.160 km
R_{major}	Earth's radius along the major axis 6378.137 km
R_{minor}	Earth's radius along the minor axis 6356.7523 km

Abbreviations

ADC	Analog-to-Digital Converter
DBF	Digital Beam Forming
FFT	Fast Fourier Transformation
HRWS SAR	High Resolution Wide Swath Synthetic Aperture Radar
IF	Intermediate Frequency
IFFT	Inverse Fast Fourier Transformation
ISLR	Integrated Side Lobe Ratio
LNA	Low Noise Amplifier
LO	Local Oscillator
MIMO	Multiple Input Multiple Output
MSE	Mean Square Error
OFDM	Orthogonal Frequency Division Multiplex
PRI	Pulse Repetition Interval
PRF	Pulse Repetition Frequency
PSLR	Peak-to-Side Lobe Ratio
Radar	RAdio Detection And Ranging
RAR	Real Aperture Radar

RF	Radio Frequency (3 kHz - 300 GHz)
RMS	Root Mean Square
SAR	Synthetic Aperture Radar
SNR	Signal-to-Noise Ratio

List of Symbols

PRF_{min}	minimal PRF value fulilling Nyquist criterion
SNR	Signal-to-Noise Ratio
SNR_i	input Signal-to-Noise Ratio of a single channel
SNR_o	output Signal-to-Noise Ratio
SNR_q	Signal to quantization Noise Ratio
j	imaginary unit, $\sqrt{-1}$
$\Re\{*\}$	real part of a complex signal
$\Im\{*\}$	imaginary part of a complex signal

Symbols with Capital Letter

A	parameter for the sliding spotlight mode
$AF(\theta)$	array factor
A_L	maximal value of parameter A in sliding spotlight mode without undersampling in azimuth
A_m	maximal amplitude, above which the ADC reaches saturation
A_p	antenna aperture size
A_F	additive factor of VKA-representation in frequency domain
B	bandwidth of the transmitted signal
B_{rg}	bandwidth of the received radar data in range
B_{az}	bandwidth of the received radar data in azimuth
$B_{overall}$	overall azimuth signal bandwidth
B_{sa}	azimuth bandwidth resulting from PRF
B_{target}	azimuth bandwidth of a single target in sliding spotlight mode

C_{const}	coefficient summarizing all constant components in radar receive signal equation
C_s	curvature factor
G	antenna gain
G_{Tx}	transmit antenna gain
G_{Rx}	receive antenna gain
G_R	characteristic on frequency domain for real part of the spectrum
G_I	characteristic on frequency domain for imaginary part of the spectrum
$I(\theta)$	antenna array pattern
$I_0(\theta)$	pattern of a single antenna element
K_s	effective FM chirp rate
K_F	factor for original signal of VKA-representation in frequency domain
N	number of independent sub-apertures in elevation
N_m	number of amplitude levels
N_{rg}	number of samples is range
M	number of independent sub-apertures in azimuth
P	number of sampling points in azimuth after azimuth filter
P_{av}	average transmit power
P_{Rx}	received power
R_{Ground}	orbit altitude over the ground
R_{Earth}	mean radius of the Earth
R_{Orbit}	orbit radius of the satellite
T	temperature in Kelvin
T_s	sampling interval
U	number of sampling points in azimuth
U_{zero}	number of additional sampling points in azimuth
V_F	factor for conjugate complex signal of VKA-representation in frequency domain

X	length of the antenna footprint on the ground
X_I	flight trajectory during the raw data acquisition
X_g	length of illuminated ground swath in the sliding spotlight mode
X_s	scene extension after azimuth filter in the sliding spotlight mode
$X(f)$	input signal in frequency domain
X_m	dynamic range
X_p	maximal amplitude of a signal
$Y(f)$	output signal in frequency domain

Symbols with Lower Case Letter

a_T	additive factor of VKA-representation in time domain
b	number of quantization bits
d_n	position of the n_{th} sub-apertures in array
dx	misplacement of antenna array element in x-plane
dx_{max}	maximal minor position error
f_{az}	azimuth frequency
f_{rg}	range frequency
f_s	sampling frequency (fast time sampling)
$f_{s,min}$	minimal sampling frequency according to Nyquist Theorem
f_c	center frequency (middle frequency)
f_{center}	frequency around which the spectrum is centered for each azimuth in sliding spotlight mode
f_D	instantaneous Doppler frequency
g_R	transmission factor of the real part of the signal
g_I	transmission factor of the imaginary part of the signal
h	altitude over the ground of the radar platform
h_{rs}	height of single receive sub-aperture
h_s	distance between two neighboring array elements in elevation
h_{rx}	height of the complete receive aperture
h_{tx}	height of the transmit aperture

$h_{t_x h_x}$	height of a single transmit/receive antenna
k_r	chirp rate
k_T	factor for original signal of VKA-representation in time domain
l	effective antenna length
l_{rs}	length of a single receive sub-aperture
l_{rx}	length of the complete receive aperture
l_{tx}	length of the transmit aperture
n_{az}	number of samples in the synthetic aperture of a target
n_i	input noise
n_{pulse}	number of range samples within one pulse
n_q	quantization noise
r	slant range
r_0	minimal distance between the radar antenna and a target
r_1	shortest distance between the steering point and the ground surface
r_{ref}	reference slant range
r_{Tx}	distance between the transmit antenna and a target
r_{Rx}	distance between the receive antenna and a target
r_s	mean slant range
$\tilde{r}$	reference range for the azimuth filter in sliding spotlight mode
$r_{me,x}$	slant range with position error in x-direction
r_{me}	slant range with position error in x-direction and in z-direction
$r_{me,z}$	slant range with position error in z-direction
s_{Tx}	transmit radar signal in baseband
s_{Rx}	receive time signal in baseband
s_n	receive time signal in baseband for n^{th} sub-aperture
t	time
t_{az}	azimuth time, slow time
t_{rg}	range time, fast time

t_n	half of the chirp time duration
t_{ref}	reference time
u	azimuth direction
u'	resampled azimuth direction
u_0	position of the target in azimuth
v_f	velocity of an antenna footprint
v_p	velocity of a plane / velocity of a radar platform
v_s	velocity of a satellite
v_s/v_f	velocity ratio
v_T	factor for conjugate complex signal of VKA-representation in time domain
w_{T_x}	transmit antenna pattern
w_{R_x}	receive antenna pattern
w_{T_x/R_x}	transmit/receive antenna pattern
x	coordinate (ground range)
$x(t)$	input signal in time
$\hat{x}[t]$	quantized input signal in time
x_A	analog signal
x_{ground}	ground distance between the projection of the radar position on the ground and the center of the illuminated swath
x_i	wanted signal
x_{me}	position error of the radar platform in x-direction
y	coordinate, (azimuth direction)
$y(t)$	output signal in time
z	coordinate, (height)
z_{me}	position error of the radar platform in z-direction

Greek Symbols

β	NaDir angle
γ_w	broadening factor of the impulse response width due to a processing window
Δ	quantization step
Δu	azimuth spacing
Δu_2	azimuth spacing after azimuth filter

$\delta(t)$	Kronecker delta
δ_{angle}	angular resolution of an radar system
δ_{az}	azimuth resolution
$\delta_{az,RAR}$	RAR azimuth resolution
δ_{ground}	ground range resolution
δ_{rg}	range resolution
θ	angle
θ_h	half width of an antenna beam
λ	wavelength at the center frequency
σ_{target}	radar cross section of a target
σ_x	RMS of the quantized signal
τ	time duration of the transmitted signal
φ_i	incidence angle
φ	illumination angle
$\Psi_n(\theta)$	phase difference in receive signal between the nth sub-aperture and the middle of the array
$\Psi_{error}(\theta)$	phase error due to element misplacement
ω_c	angular center frequency
ω_f	circular velocity of the antenna footprint
ω_s	circular velocity of the satellite

1 Introduction

The word 'radar' comes from a phrase 'RAdio Detection And Ranging'. It was devised by the United States Navy in November 1940, and officially adopted by the Allied Powers in 1943, [87]. A radar is a device that transmits an electromagnetic wave, which is reflected on a target and received again by the radar. The range, altitude, direction and speed of the illuminated object can be identified with the information carried by the received wave by adequate processing of the received signal.

The idea of detecting objects with the help of electromagnetic waves has its grounds in an experiment of Heinrich Hertz that took place in 1887/1888 in Karlsruhe, [10]. This experiment provided the proof that transverse free space electromagnetic waves can travel over some distance. It gave the basis for all wireless radio communication and radar systems. The idea of a radar itself was first presented by Nikola Tesla in an article published in the *Century Illustrated Monthly Magazine* in June 1900. In 1904 German engineer Christian Hülsmeyer patented a method of detecting distant metal objects with electromagnetic waves, [41], [96]. His device called *Telemobiloskop* could detect ships from 3000-meter distance. The device could detect the presence of a ship also in dense fog.

A strong impulse on the emerging of radars of all kinds were the First and the Second World War. The initial purpose of radar systems was to detect and track aircrafts and ships through heavy weather and darkness. The original radar system measured the direction and the range to a target via antenna directivity and time delay of the transmitted pulse accordingly. One of the most famous radar systems during the World War II was the ring of coastal Early Warning radar stations build by the British with the code name Chain Home. The system provided a long-range detection of aircrafts through radars fixed on top of radio masts. The system was based on the patent of Robert Watson-Watt from 1935. During the war the rapid development of radar technology was concentrating on military application on land, sea and air. After the war radar widened its application fields including for example also civil aviation, marine navigation, medicine, meteorology [11] and traffic control.

In 1951 Carl Wiley from the Goodyear Aircraft Corporation patented an alternative means of improving azimuth resolution of a radar system. This concept was called *Synthetic Aperture Radar (SAR)* to signify that a signal synthesis was needed to receive a radar image. The idea is to synthesize a very long antenna aperture via signal processing. The general concept of SAR is to carry a single, relatively small, antenna on an aircraft and combine return signals by the means of signal processing to produce a high-resolution image of the terrain comparable to what could be obtained by a much larger and more expensive antenna. The amount of data processing required was too high for hardware existing at that time. The optical processing of SAR data was the primary processing technique for the next three decades [21]. A SAR system provides a two-dimensional image of the illuminated area, which can be easily compared with optical images and has its application mainly in mapping and remote sensing. The main advantage of a radar over an optical system is the independence on the sun illumination and weather conditions. Electromagnetic waves of common radar frequencies pass through clouds and precipitation with little or no deterioration. In the following years more complex systems and processing algorithms were proposed and implemented.

A very important part of the SAR system is the signal processing algorithm also called SAR processor. Unlike other types of radars the synthetic aperture radar requires a complex signal processor before data can be evaluated. The raw data obtained directly from the SAR system looks for a user much like noise and needs to be processed with an adequate SAR algorithm. The first real-time digital SAR processor built by MacDonald Dettwiler was delivered to CCRS (Canada Centre for Remote Sensing) in 1979, and installed on the Convair-580, [12]. Since that time SAR processing algorithms have become more complex and precise. Modern digital electronics with rapidly raising computational capabilities made the development of more complex algorithms for SAR processing possible and by this increase the SAR system performance.

1.1 Synthetic Aperture Radar

The SAR system can be mounted on various platforms, e.g. an airplane, a helicopter or a satellite. If the radar platform is an airplane or similar the system is commonly called an airborne SAR system. A spaceborne SAR system is mounted on a satellite. The principle is the same for all types of SAR systems independent of the platform. The synthetic aperture radar system consists of two important components: one is the radar system hardware, the other is the SAR processing algorithm. In order to get the good quality radar image both components must fulfill necessary requirements.

The first successful focused airborne SAR image was provided in August 1957. Instruments providing demanded signal linearity and frequency stability were designed and built by the Radar Laboratory. The AC–46 (Curtiss Commando) aircraft was used as a radar platform. The 5–foot (1.5 m) antenna used for illumination was theoretically capable of 5–foot (1.5 m) resolution. The goal was although to demonstrate 50–foot (15 m) resolution, as only 10% of the beam width was used at first. Finer resolution would require correction of non ideal flight path, which was not available at this time. Still over numerous trials even small turbulences kept the aircraft from flying straight and level enough making the data insufficient for processing. Finally in August 1957 a map-like image of the Willow Run Airport, which demonstrated required resolution in some parts of the image, was obtained.

It was understood that a satellite out–of–atmosphere platform with an extremely smooth orbital path is suited for SAR application. The first civilian spaceborne SAR, which was designed for remote sensing of the Earth's oceans was the SEASAT launched in June 1978, [57]. The SEASAT was working in the L-Band with HH polarization. It was developed and managed by the Jet Propulsion Laboratory (JPL). With 19 MHz bandwidth a resolution of about 25 meters in azimuth and range has been achieved. Most of the data was processed optically, where the 100 km of the whole swath length was divided into four 25 km long parts. The SEASAT was followed by the Shuttle Imaging Radar–A (SIR–A) and the Shuttle Imaging Radar–B (SIR–B) in 1981 and 1984, accordingly [34]. Both radars operated at the L–band with HH polarization. A significant increase in the number of SAR missions could be seen in 1990s. In July 1991 the first European Earth–observing satellite ERS-1 (European Remote Sensing) was launched. The satellite was equipped with, among other, instruments with a synthetic aperture radar working in the C–band. A successor mission ERS–2 was launched in April 1995. In 1992 the Japanese Earth Resources Satellite 1 (JERS–1) was launched equipped with a L–band SAR.

The RADARSAT–1 launched in November 1995 by Canadian Space Agency (CSA) could be operated in different modes: fine, which had a coverage of

50 km by 50 km with resolution of 10 m, standard with coverage of 100 km by 100 km with resolution of 30 m and scanSAR with large coverage of 500 km by 500 km with resolution of 100 m. The successor RADARSAT–2 satellite was launched in December 2007. The highest resolution in the ultra fine mode of this SAR system is 3 m [2].

The Shuttle Imaging Radar–C/X–band SAR (SIR–C/XSAR) joint NASA, German Space Agency (DARA) and Italian Space Agency (ASI) mission have been flown in April and October 1994 [36]. The system could be operated simultaneously at three frequencies (L, C and X–band). The radar system had also the ability to transmit and receive at both horizontal and vertical polarizations, providing "fully polarimetric" capabilities.

The Shuttle Radar Topography Mission (SRTM) obtained elevation data on a near-global scale during an 11–day mission in February 2000 [74]. The data were used to generate the most complete high–resolution digital topographic database of the Earth. The SRTM provided a single–pass interferometry with two radar receive antennas, one located in the Shuttle's payload bay and the second receive antenna positioned on the end of a 60–meter mast that extended perpendicular to the main radar antenna, once the Shuttle was in space, [1].

The ENVISAT (Environmental Satellite) was launched in March 2002 [20]. This European Space Agency (ESA) Earth–observing satellite with an ASAR (Advanced Synthetic Aperture Radar) sensor operating at the C–band added the capability to collect data with choice of five polarization modes: VV, HH, VV/HH, HV/HH, or VH/VV. The contact with the ENVISAT was lost on 8nd April 2012, and on 9nd May ESA formally announced the end of the mission.

The TerraSAR–X is a German Earth observing satellite launched in June 2007 as a joint venture being carried out under a public-private-partnership between German Aerospace Center (DLR) and EADS Astrium GmbH [6], [56]. As the name suggests this radar is working in the X–band, on a center frequency of 9.65 GHz. The radar platform revolves the Earth in a polar orbit at an altitude of 514 km. The TerraSAR–X acquires new-quality images with up to 1 m resolution of the entire Earth's surface [3]. The data can be acquired in stripmap, spotlight, sliding spotlight or scanSAR mode depending on required parameters of the radar image. Another very interesting mission is the launch of a second, very similar to TerraSAR–X satellite in June 2010. The name of this mission is the TanDEM–X TerraSAR–X add-on for Digital Elevation

Measurements. Both satellites form a unique constellation, which allows the generation of global digital elevation models (DEMs) at an unprecedented accuracy, coverage and quality. The system provides high quality data with across-track and along-track interferometry as well as new techniques with bi-static SAR like polarimetric SAR interferometry [55].

The first German SAR based reconnaissance satellite system is the SAR–Lupe, where the name refers to the German word for magnifying glass. This system consists of five satellites. The first one was launched on 19$^{\rm nd}$ December 2006 [30]. The following satellites were launched in about six–months intervals until the last launch on 22$^{\rm nd}$ July 2008. At this time the system has achieved full operational readiness. The SAR–Lupe works in the X–band. The constellation of the five satellites is chosen such so that the illumination areas of the following satellites close without gaps [7], which gives full coverage of Earth's surface.

The CyroSat launched successfully in April 2010 acquires measurements of the thickness of the Earth's continental ice sheets and marine ice cover [19]. The planned duration of the mission is minimum three years. The primary payload of the CyroSat is the SAR/Interferometric Radar Altimeter (SIRAL).

The SAR sensor can be used for many various applications depending on features of the system like for e.g. resolution, polarimetry and interferometry techniques or platform (airborne, spaceborne) specifics. The spaceborne SAR systems are used mostly for topographic mapping, land cover, land usage mapping, biomass mapping and traffic monitoring [1]. A SAR can be also used to monitor the ice cover, changes in ground surface like for e.g. volcano eruption, agriculture and vegetation monitoring, water monitoring, ship tracking [92], ocean wave monitoring [80], reconnaissance mission, imaging of archaeological sites or sea shore monitoring. The TerraSAR–X for e.g. provides a rapid emergency response in case of natural or man-made disasters. The service was deployed e.g. to estimate the scale of the flood in England in July 2007 [81], or the eruption of the Eyjafjallajoekull volcano in April 2010, see Fig. 1.1.

SAR systems are used not only to map Earth but also for mapping other planets or moons. The Magellan spacecraft launched by NASA in May 1989 arrived at Venus on August 1990 [15], [35]. The primary objective of the four year mission was to map the surface of Venus with a synthetic aperture radar and to determine the topographic relief of the planet. During this mission 98 % of the Venus surface was imaged at resolution better than 100 m.

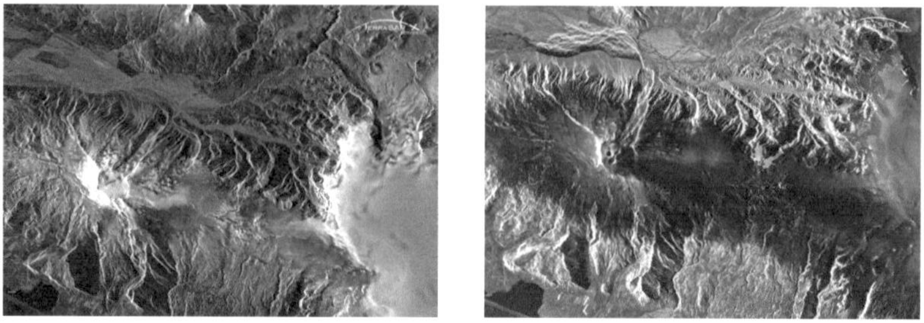

(a) before explosion (b) after explosion

Figure 1.1: SAR images of the Eyjafjallajoekull Volcano from April 2010. Source: Infoterra GmbH 2011, http://www.infoterra.de

1.1.1 Future Trends

Main demands on future SAR systems are higher resolution, larger scene size, faster acquisition of images, delivery and a possibility of obtaining SAR images of the same area in the shortest possible time as required e.g. for disaster recovery. This reflects the general trend in the imaging radar development to provide better quality images with higher resolution, wider coverage and shorter revisit times [47]. Future systems tend also to be more flexible with various operating modes obtaining images fitting specified requirements [103]. Multi-channel SAR systems will overcome the limitation of contradicting requirement for wide unambiguous swath coverage and high azimuth resolution [44], [45]. The full DBF improving the agility and the performance of the resulting radar image is in the focus for future military and commercial applications, [33], [101]. The radar interferometry and polarimetry provide new information on the illuminated ground area [77]. More interest is also put to satellites with a SAR system on board flying in a constellation. The successor mission of the ENVISAT is the Sentinel-1 [18]. It is a two satellite constellation with a prime objective of land and ocean monitoring. The two satellites are scheduled to be launched for 2013 and 2015.
The main focus of future system is slightly different for civil and for military applications [86]. The main demand for civil SAR systems is to cover a large area with a singe measurement. The measurement in some applications like e.g. ice cover monitoring is repeated several times to record changes. Relatively short revisit times and good resolution are of importance. The

military mission requirements are a very high resolution image of a relatively small area within a very short time and possibility of multiple acquisitions of the same area.

1.2 Motivation and Objective of this Work

The objective of this work are various aspects of a high resolution SAR system. The synthetic aperture radar system consists of two important components: one is the radar hardware, and other is the SAR processing algorithm. Both components are crucial for system performance. Two radar system concepts, which use a beam steering to improve performance are considered. The first system is the high-resolution wide-swath synthetic aperture radar (HRWS SAR). The HRWS SAR loosens the dependence between the high azimuth resolution and wide range coverage through means of digital beam forming (DBF) in both elevation and azimuth direction. This allows for an improvement in resolution of the radar image and at the same time large coverage range. In the HRWS SAR the beam is steered electrically with an antenna array. The system consists of multiple independent receive paths, where each element of the receive aperture has its own front-end.
The second concept within the scope of this work is the sliding spotlight mode. In this mode the radar antenna beam is slided over the Earth's surface. The sliding of the beam can be achieved through mechanical or electrical beam steering or rotation of the whole satellite. The sliding spotlight mode provides a good compromise between high resolution and wide coverage. Improving resolution of the radar system puts more pressure on accurate processing algorithms, which must handle highly resolved data.

1.3 Scope of this Work

The performance of a SAR system depends on the quality of the hardware used for recording the data and the accuracy of the processing algorithm. The first part of the this work considers aspects of the radar hardware. The nonidealities in the receiver can reduce the performance of the whole SAR system. The same applies to antenna array. A misplacement of antenna array elements will influence the quality of the radar image. The quantization is also an important factor. The dependence of the quality of the radar image on the number of quantization bit is investigated. The second part of the work considers the processing algorithm. The inaccuracies in the processing will reduce the quality of the image. To achieve a high resolution radar image all the components of hardware and signal processing must provide good quality.

The structure of this work is as following. Basics of the synthetic aperture radar are described in Chapter 2. The generic SAR processing is described and main SAR modes are presented. The principle of the DBF in azimuth and in range is explained shortly. The point target and parameters describing radar image quality are defined. Those parameters are used in following chapters.

Following chapters of this work can be divided into two main parts. Design aspects of a multi-channel synthetic aperture radar with DBF on receive are considered in the first part. An example of such radar system is the HRWS SAR. The resolution improvement of radar images requires the transmit and receive front-end with high performance. The influence of various nonidealities of the receive path is considered in Chapter 3. The direct down-conversion receiver is used as a model for the receiver simulation. The same type of the receiver was used in the HRWS SAR demonstrator that was developed and built by EADS Astrium GmbH. Components of this demonstrator receiver were measured to provide realistic parameters for the simulation. The simulation showing the influence of the receive front-end on the performance of the radar was developed. A new representation describing receiver components and possible effects in a receiver is proposed. This representation can model receiver components inclusive possible nonidealities in the system. It can be easily implemented into the simulation and is efficient in computation.

A characteristic feature of the HRWS SAR is a very large receive antenna array. A possible problem of such system is a bend of the antenna plane. The influence of a single panel of the receive antenna array in elevation is described in Chapter 4. Two possible bends of antenna plane are considered. The influence of those minor position errors on the array factor is simulated. The HRWS SAR simulation used for calculating effects of the receiver front-end on the radar signal is extended to include the potential position error for each array element.

The idea behind HRWS SAR is to improve the performance of the system with the DBF in azimuth and in elevation. The system has multiple independent receive channels. Each of the channels provides a receive signal that has to be saved and/or processed. Using N independent receive antennas increases the amount of collected raw data N times. Low bit quantization is an idea to reduce the amount of the radar raw data to be recorded and/or processed with a SAR compression algorithm. The radar data, that have been quantized with only one or two bits, can be reconstructed by the means of the DBF in range. The working principle of low bit quantization is described in detail in Chapter 5. Special interest is put on the quantization with a single bit. In this case a simple comparator can replace the Analog-to-Digital Converter (ADC). The difference between the quantization with multiple bits and low bit is investigated by the means of simulation. The difference between the quantization of signals with low or high additive white noise is explained. The HRWS SAR demonstrator has eight independent receive channels with one bit quantization. A measurement of the demonstrator illustrates the principle of the low bit quantization and the signal reconstruction through the DBF processing. The simulation used for previous chapters was extended to include the ADC block with an adjustable number of quantization bits. The optimal quantization parameters: the number of quantization bits, the signal to noise ratio of the receive radar signal and the number of independent receive channels are determined with help of the simulation. The concept of the low bit quantization is evaluated for its possible application in the SAR system.

The second part of this work considers a special illumination mode that uses e.g. the rotation of a satellite to slide the antenna beam over the illuminated ground and thus increase the size of the illuminated area. This special illumination mode is called sliding spotlight or hybrid strip-map/spotlight mode. A compromise between the azimuth resolution and the swath coverage is achieved through a beam steering in azimuth. An adequate processing algorithm is necessary to compress the raw radar data. An airborne SAR radar working in the sliding spotlight mode is described in Chapter 6. The two proposed processing algorithms process the sliding spotlight raw data. One is based on the chirp scaling algorithm and the second one is based on omega-K algorithm. Both algorithms are modified to cope with the sliding spotlight mode raw data. A simulation for the sliding spotlight mode is developed. The simulation consists of two main modules: the generation of raw data and the SAR processing algorithm. The performance of both processing algorithms is evaluated for various range and azimuth resolutions.

In Chapter 7 proposed processing algorithms are modified to cope with the spaceborne SAR raw data. The algorithm must be adjusted to the trajectory of the satellite. First algorithms are adjusted for an ideal round satellite orbit. A simulation for the spaceborne SAR geometry is developed. The simulation generated the radar raw data for a specified scenario. The raw data can be then processed with an adequate compression algorithm. Both proposed algorithms are evaluated for various range and azimuth resolutions. Then processing algorithms are modified to cope with a real satellite trajectory also with possible motion errors. An adequate correction block for a real trajectory is proposed for both processors based on the chirp scaling and on the omega-K algorithm. The real orbit correction is evaluated by the means of simulation.

2 Principle of Radar with Synthetic Aperture

In this chapter fundamentals of a synthetic aperture radar (SAR) are presented. The chapter begins with the radar equation describing the relationship between transmit and receive signal power for all types of radar systems. The SAR system geometry, working principle, generation of transmit and receive signals and an overview of general SAR signal processing methods is shortly described. Next, various SAR modes and basic digital beam-forming (DBF) algorithms in elevation and in azimuth direction are described. The last section of this chapter characterizes an ideal point target and defines various figures of merit for quality of radar images, which are used in this work.

2.1 Radar Equation and Radar Principle

A radar is a system that transmits an electromagnetic wave and receives the echo reflected on a target. Let's assume that the radar system is transmitting a signal with an average transmit power of P_{av}. The radar transmit antenna has a gain G_{Tx}. The transmitted signal has a linear frequency modulation (chirp signal) with a carrier wavelength of λ and a bandwidth of B. The radar signal is reflected on a target with a radar cross section of σ_{target} positioned at a distance r_{Tx} from the transmit antenna. The signal reflected on the target is received by the radar receive antenna with a gain of G_{Rx}. In general case the radar system has a separate transmit and receive antenna. The receive antenna is positioned at a distance r_{Rx} from the target. The received power P_{Rx} can be calculated according to the radar equation, Eq. (2.1)

$$P_{Rx} = P_{av} \frac{G_{Tx} G_{Rx} \lambda^2}{(4\pi)^3 r_{Tx}^2 r_{Rx}^2} \sigma_{target} \tag{2.1}$$

The radar equation can be simplified for a monostatic radar with a single antenna for transmit and receive. The gain for the transmit and receive antenna

$G_{Tx} = G_{Rx} = G$ is then the same. The distance between a target and the transmit/receive antenna $r_{Tx} = r_{Rx}$ is also the same.
A very important parameter characterizing a radar system is its resolution. The angular resolution for a real aperture radar (RAR) depends on the half width of the antenna beam θ_h and the distance between the radar antenna and the illuminated target. In order to improve the resolution of a RAR the directivity of the antenna must be enhanced. This can be achieved by reducing the wavelength of the transmitted signal or by increasing the length of the antenna. Improvement of the azimuth resolution can be also achieved through the principle of synthetic aperture.

2.2 SAR Fundamentals

In the synthetic aperture radar a longer antenna aperture is synthesized by coherent combination of receive signals [8], [12], [13]. A large synthetic aperture improves the resolution of the radar system in azimuth direction without an actual increase of the antenna length. The radar antenna is mounted on a moving platform, which can be e.g. an airplane for an airborne SAR or a satellite for a spaceborne SAR. Fig. 2.1 shows the geometry of an airborne SAR system. The airplane is moving on a straight trajectory with a constant velocity v_p at a defined height h. The direction, in which the platform is moving, is referred to as the azimuth direction. The illuminated ground swath in Fig. 2.1 is marked green. The resulting SAR image is a two-dimensional complex matrix. The two dimensions of a SAR image are azimuth (in flight direction) and range (perpendicular to the flight direction). The range can be also defined along the direction the radar is pointing at and called slant range. The ground range is a projection of the slant range on the ground and can be easily recalculated if the geometry of the radar system is known. The ground range and the slant range are both marked in Fig. 2.2. Further in this work the term range direction will refer to the slant range direction and range resolution will refer to resolution of the system in slant range.
The radar antenna in Fig. 2.1 is positioned perpendicular to the flight direction. In the general case the radar antenna can be also placed skewed against the azimuth direction by a squint angle[1]. In this work for all considered sys-

[1] The squint angle is an angle between the airplane perpendicular to the flight direction of the radar platform (aircraft or satellite) and the beam direction. This angle, if not zero, must be

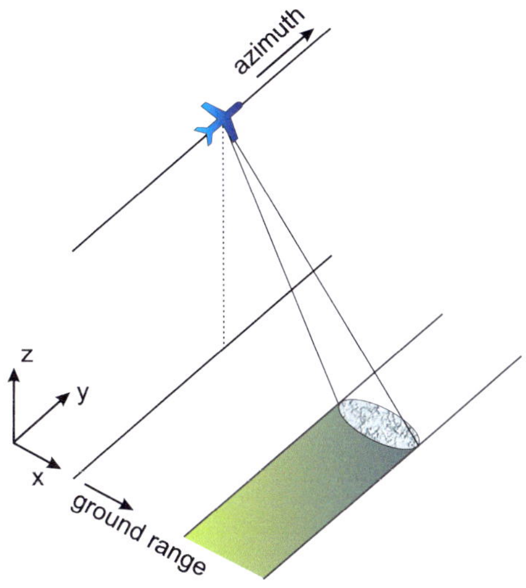

Figure 2.1: The geometry of an airborne SAR system.

tems the squint angle is zero. A SAR system is a Side Looking Airborne Radar (SLAR), which means that the radar antenna is positioned in such way, that the ground is illuminated on one side of the platform – left or right side. The angle under which the SAR antenna is positioned in elevation is called NaDir angle and is described by the Greek letter β, see Fig. 2.2. The angle between the incident wave emitted by the radar antenna and the normal to the ground is called incident angle φ_i. Both angles are marked for an airborne SAR in Fig. 2.2. In the airborne case both angles can be assumed equal as the Earth curvature can be neglected. For a spaceborne SAR the difference between NaDir angle β and incident angle φ_i is significant for SAR processing and must be therefore considered in calculations.

2.2.1 SAR Transmit Signal

The radar platform moves forward while transmitting signals in specified time intervals. These intervals are chosen so that no ambiguities occur in the SAR image as well as transmitting and receiving times do not overlap. The time

taken into consideration in SAR processing.

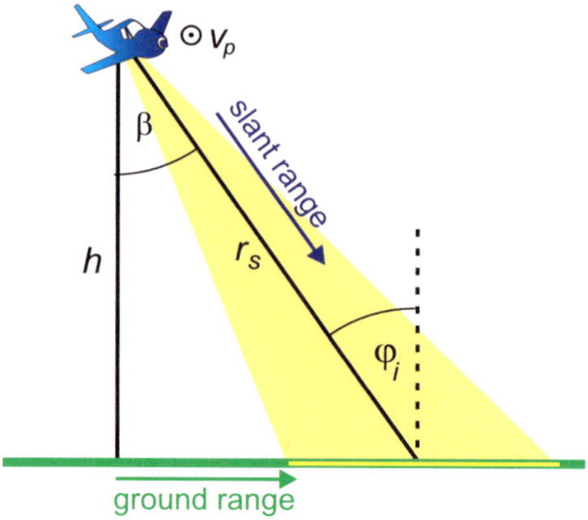

Figure 2.2: The geometry of an airborne SAR system a cut-view through the range plane

interval between transmission of two signals is called Pulse Repetition Interval (PRI) or more often it is recalculated to a frequency and called Pulse Repetition Frequency (PRF)[2]. The transmit signal in SAR systems is usually a linear chirp signal as described with Eq. (2.2). The instantaneous frequency of the transmitted signal is changing linearly with time t.

$$s_{Tx}(t) = C\exp(j\pi k_r t^2) \cdot \text{rect}\left\{\frac{t - t_n}{\tau}\right\} \tag{2.2}$$

τ is the duration of the chirp signal and t_n is defined as half of chirp duration. k_r is the linear chirp rate defined according to Eq. (2.3). C is the amplitude of the transmitted signal.

$$k_r = \pm\frac{B}{\tau} \tag{2.3}$$

where B is the bandwidth of the transmitted signal. The minimal sampling frequency for the baseband complex valued signal with bandwidth B according to the Nyquist Theorem is then:

$$f_{s,min} = B \tag{2.4}$$

[2] PRF and PRI can be easily recalculated: $\text{PRF} = \frac{1}{\text{PRI}}$

According to Eq. (2.3) the chirp rate can be positive or negative. In case k_r is negative the signal in Eq. (2.2) is a down-chirp, where the instantaneous frequency decreases continuously with time. If k_r is positive the resulting signal is an up-chirp, where the instantaneous frequency increases with time. In further chapters an up-chirp with positive k_r value is assumed. The signal processing is the same independent on up- or down-chirp being used.

2.2.2 SAR Receive Signal Model

Eq. (2.2) shows the radar transmit signal in base band. For transmission the signal is up converted to the carrier frequency f_c. The up converted chirp signal is transmitted by the radar antenna. The antenna receives the signal reflected on various targets. Let's assume a single point target positioned at a minimal distance of r_0 from the radar antenna. The received signal is down converted back to the baseband. The receive and down converted signal $s_{Rx}(t)$ is described with Eq. (2.5).

$$\begin{aligned} s_{Rx}(t) &= C_{const} \cdot \sigma_{target} \cdot w_{Tx/Rx}(t) \cdot \ldots \\ & \exp\left(-j\pi k_r \left(t - t_n - \frac{2r_{target}}{c_0}\right)^2\right) \cdot \ldots \\ & \text{rect}\left\{\frac{t - t_n}{\tau}\right\} \end{aligned} \tag{2.5}$$

c_0 is the speed of light in vacuum, $w_{Tx/Rx}$ is the antenna beam pattern of the transmit/receive antenna. The time dependency of the antenna pattern indicates that the direction of the incoming receive radar signal is changing with time due to the system geometry. The coefficient C_{const} summarizes all constant factors of the equation such as transmit power, free-space attenuation and influence of the transmit and receive paths. In further analysis the antenna patterns and the coefficient C_{const} are omitted, if not explicitly necessary, as they have no influence on described signal processing algorithms.

The SAR system during swath illumination and data collection moves with a constant velocity v_p and transmits a radar signal in constant time intervals. The minimal required PRF is the minimal sampling frequency in azimuth direction fulfilling the Nyquist Theorem. In order to provide an unambiguous SAR image the swath should be sampled with a distance between two samples not larger than half of the effective antenna length $l/2$. The minimal PRF for

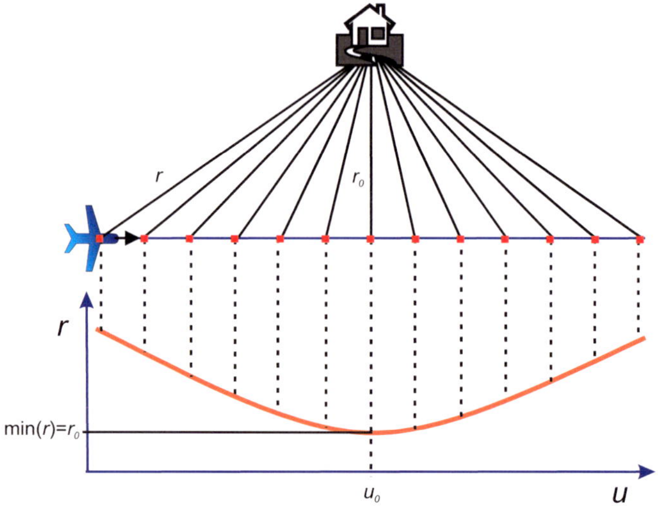

Figure 2.3: The change of the distance between the radar antenna and the target

a specified antenna and platform velocity is described with Eq. (2.6).

$$\mathrm{PRF}_{min} = \frac{2v_p}{l} \tag{2.6}$$

The distance between the target and the radar antenna r is changing during the flight. The minimal distance is called r_0 and describes the position of the target in range. Eq. (2.7) shows the distance in dependence on flight time. In Fig. 2.3 explained is the change in the distance from the radar antenna to the target. The plane with radar on-board is flying on a straight line over the illuminated target (the house). Assumed is a target positioned at (r_0, u_0), where r_0 is the shortest distance between the radar antenna and the target, u_0 is the position of the target in azimuth. u is the azimuth direction. At the positions marked with red dots the radar takes measurements. The distance between the radar and the target r for each position of the plane is presented underneath. The shortest distance r_0 is when the plane is above the target at the same position in azimuth u_0. The range curve is a parable.
The distance between the target and the radar antenna r in dependence on time

can be easily calculated from the geometry of the radar system as in Eq. (2.7). It is assumed that the radar antenna is stationary during transmission and reception of a single chirp signal and then moves to the next azimuth position, where the next radar signal is transmitted.

$$r(t) = \sqrt{r_0^2 + (u(t) - u_0)^2} = \sqrt{r_0^2 + (v_p t - u_0)^2} \tag{2.7}$$

To describe the dimensions of a two dimensional SAR image used is fast-time, also called range time t_{rg}, and slow-time also called azimuth time t_{az}. The fast-time can be easily recalculated from slant range r_s as shown in Eq. (2.8), see also Fig. 2.2. The slow-time can be defined through the position of the radar platform in azimuth direction u assuming constant monotonous motion of the radar platform. This relationship is described with Eq. (2.9).

$$r_s = \frac{c_0 t_{rg}}{2} \tag{2.8}$$

$$t_{az} = \frac{u}{v_p} \tag{2.9}$$

The two dimensional SAR receive signal is described with Eq. (2.10).

$$\begin{aligned} s_{Rx}(t_{rg}, u) = {} & \exp\left(-j2\pi f_c \frac{2r(u)}{c_0}\right) \cdot \ldots \\ & \exp\left(-j\pi k_r \left(t_{rg} - t_n - \frac{2r(u)}{c_0}\right)^2\right) \cdot \ldots \\ & \mathrm{rect}\left\{\frac{t_{rg} - t_n - \frac{2r(u)}{c_0}}{\tau}\right\} \mathrm{rect}\left\{\frac{u - u_0}{X}\right\} \end{aligned} \tag{2.10}$$

Where X is length of the antenna footprint on the ground corresponding to the synthetic aperture length. It can be calculated according to Eq. (2.11), [12].

$$X = \frac{0.886\lambda}{l} r_s \tag{2.11}$$

The synthetic aperture depend on the effective antenna length l and the middle slant range r_s. The factor 0.886 comes from the antenna beam width calculation [12]. The length of the synthetic aperture corresponds to the projection of the 3 dB beam width of the radar antenna beam on the ground. X is slightly changing over range. Usually the value estimated for the middle of the illuminated swath is considered to characterize the radar system. The window in azimuth direction during which the target is 'seen' by the radar antenna is calculated with Eq. (2.12). The window extension depends on the 3 dB band width of the radar antenna and the movement (platform velocity) in azimuth direction. A target positioned at azimuth u_0 can be seen by the radar antenna as long as it is illuminated with antenna beam.

$$|u - u_0| \leq \frac{X}{2} \tag{2.12}$$

Eq. (2.10) describes a two dimensional SAR receive signal from a single point target. A complex scene with many various targets can be seen as a superposition of responses from all targets within the illuminated scene.

2.2.3 Resolution in Range and Azimuth

An important measure of the performance and the quality of a SAR system is the resolution in range and in azimuth direction. The resolution in range depends on the bandwidth of the transmitted chirp signal. The term range resolution is used here for slant range resolution. The ground range resolution can be recalculated from the slant range resolution with known incidence angle. The range resolution can be calculated according to Eq. (2.13). γ_w is the broadening factor of the impulse response width due to a window used in processing algorithm. This factor is often omitted as it depends on the processing chain and not on system parameters.

$$\delta_{rg} = \frac{c_0}{2B}\gamma_w \approx \frac{c_0}{2B} \tag{2.13}$$

The ground range resolution, as its name suggests, considers the resolution of the radar image mapped onto the illuminated surface. The slant range resolution is constant over the whole swath width of the radar image. The ground range resolution changes slightly with the illumination angle φ over the swath. The incidence angle φ_i is the illumination angle measured to the middle of the swath. The simple way to calculate ground range resolution from slant range resolution is according to Eq. (2.14).

$$\delta_{ground} = \delta_{rg}\sin(\varphi) = \frac{c_0}{2B}\gamma_w \sin(\varphi) \tag{2.14}$$

The azimuth resolution is defined as the minimal distance in azimuth between two objects on the ground at which both targets can be distinguished. The distance equal to the 3 dB beam width projection of the antenna pattern on the ground is assumed to be the measure of resolution. The resolution of a real aperture radar (RAR) system is expressed as in Eq. (2.15).

$$\delta_{az,RAR} = r\theta_h \gamma_w = \frac{0.886\lambda r}{l}\gamma_w \tag{2.15}$$

θ_h is the half width of antenna defined as in Eq. 2.16. The factor 0.886 is related to the estimation of the antenna beam width according to Eq. (2.16), [12].

$$\theta_h = \frac{0.886\lambda}{l} \tag{2.16}$$

It should be noted that the azimuth resolution for a RAR system depends also on the position of the target in range r. For a SAR system a large antenna aperture is created as the radar platform moves over the illuminated area. The synthetic aperture is slightly longer for a far target and shorter for a near target. Usually the synthetic aperture is calculated for the slant range to the middle of the illuminated swath r_s. The resolution for the SAR system in strip-map mode[3] (see Section 2.4.1) is described with Eq. (2.17)

$$\delta_{az} = \frac{0.886\lambda r}{2X}\gamma_w = \frac{0.886\lambda r}{2\frac{0.886\lambda r}{l}}\gamma_w = \frac{l}{2}\gamma_w \tag{2.17}$$

Usually the broadening factor of the impulse response width due to a processing window is not considered. The Eq. 2.17 is then reduced to:

$$\delta_{az} = \frac{l}{2}. \tag{2.18}$$

The factor 2 in the denominator in Eq. (2.17) is a consequence of the geometry of the SAR system. In RAR the signal is transmitted and received at almost the same time and the same azimuth antenna position. The synthetic aperture in SAR is created upon data collected in succession at separate positions. In this case the distance from the radar antenna to the illuminated target and also the way back counts into the phase change of the received signal. The phase

[3] For other SAR modes the azimuth resolution calculation must also consider the influence of antenna beam steering on azimuth resolution.

difference for signals received on different positions of the synthetic aperture is twice as large as for the real aperture.
To notice is that the azimuth resolution in SAR depends only on the effective length of the radar antenna l. It does not depend on the distance to the illuminated scene.

2.3 General SAR Signal Processing

The special property of a SAR system compared to a pulse radar is a sophisticated signal processing necessary to access the information concealed in the receive raw data. The receive raw SAR signal looks much like random noise and it is mostly not possible (except for some very strong point targets) to recognize targets and objects before applying an adequate signal processing. A response from a single point target, see Eq. (2.10), is spread over many receive samples in both range and azimuth direction. Fig. 2.4 shows a general SAR signal processing chain. The first image on the left in Fig. 2.4 shows the phase of the complex raw data for a single point target scenario. It should be noted that the received energy from a point target positioned in the middle of the scene for the uncompressed data is spread over the whole example image. In order to get a radar image the raw SAR data must be adequately processed. For this purpose, various processing algorithms have been developed and are still being further improved. In generic SAR compression algorithm three general main processing steps can be distinguished: range compression, range cell migration correction (RCMC) and azimuth compression. The general block diagram of the processing algorithm is presented in Fig. 2.4 with all three main steps. For each step an image of a single point target in the middle of the illuminated scene is presented. Each image shows how the processed data looks at this stage of the processing.

2.3.1 Range Processing

The first step in the general SAR processing is the range compression. The radar transmits a chirp signal, according to Eq. (2.2), which is extended in time. In pulse radar the distance between the radar antenna and the target is calculated directly from the receive signal time delay. To extract the precise position in range from the SAR received signal the data must be compressed.

This can be done through matched filtering by convolution with the complex conjugate transmit signal. The convolution in the time domain requires much computing power and is therefore time consuming. Because of this the convolution is usually implemented as a multiplication in the frequency domain. The second image from the left in Fig. 2.4 shows the SAR data after range compression. In this image the compressed in range data is "squeezed" to a thin line, where most energy concentrates on few range bins (see Fig. (2.4)).

2.3.2 Range Cell Migration Correction

The second step in generic SAR processing is the range cell migration correction (RCMC). This step can be sometimes omitted in case of low quality low resolution SAR images. The distance between the target and the radar antenna changes while the radar is moving in azimuth direction illuminating the target, see Fig. 2.3. This change is described by Eq. (2.7). The receive signal from a target "migrates" in range in the raw data. This effect can be clearly seen in Fig. 2.4 in the second image. The range compressed receive data of a single target forms a parable. This effect is called Range Cell Migration (RCM). To mention is that the bending of the receive signal depends on the minimal distance to the target, incidence angle and squint angle. Near targets will show stronger bent in the range compressed data as far targets. Before the data can be processed in azimuth this effect is to be corrected. This may be achieved in many different ways, e.g., through interpolation or through shifting with phase multiplication. Arranging receive range compressed data in a straight line is important for the following azimuth compression.

2.3.3 Azimuth Processing

The last processing step is the azimuth compression. The matched filter for the azimuth direction can be calculated from the geometry of the radar system. As the SAR system moves creating the synthetic aperture, the phase of the receive signal is changing accordingly. Resulting is a modulation of the receive signal in azimuth direction, which is similar to a chirp modulation. A fully compressed radar image is provided after this step. The magnitude of the fully compressed image of a point target can be seen in the last image on the right in Fig. 2.4.

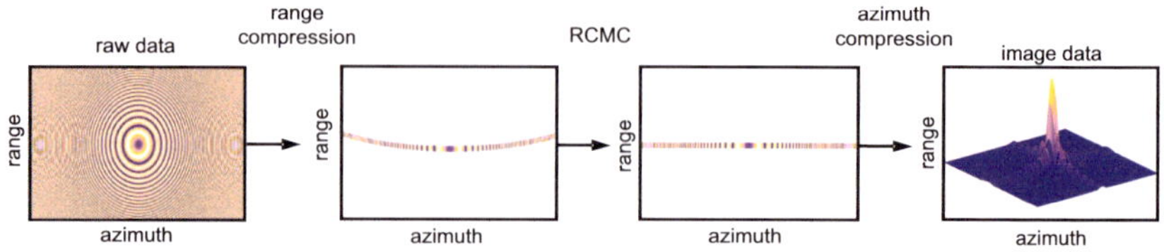

Figure 2.4: The general signal processing chain of the SAR raw data.

2.4 SAR Modes

A SAR system can be operated in various modes. The simplest acquisition mode is the strip-map mode, where the radar antenna beam is fixed. For other modes the radar antenna can be steered in both azimuth and elevation direction. The steering can be realized mechanically or electrically. Through this steering higher azimuth resolution, longer or wider swath coverage can be achieved. In the following paragraphs most common SAR modes are briefly described.

2.4.1 Strip-Map Mode

The simplest and most often used mode is the strip-map mode. In this mode the radar antenna beam is fixed, so that the continuous illumination of a long swath is possible. The geometry of this mode is presented in Fig. 2.5(a). The antenna footprint on the ground for an airborne SAR moves with the same velocity v_f as the radar platform v_p. The radar antenna can be squinted in the plane perpendicular to the direction of the platform's motion by a constant squint angle. In the example system presented in Fig. 2.5(a) the squint angle is equal to zero. The advantage of the strip-map mode is a continuous illumination of the swath in azimuth direction. The length of the swath is not explicitly limited by the radar geometry. The disadvantage of this mode is the limitation of the azimuth resolution to half the radar antenna length.

2.4.2 Spotlight Mode

In the spotlight mode the radar system is optimized in regards to azimuth resolution. The antenna beam is steered in azimuth direction to illuminate the

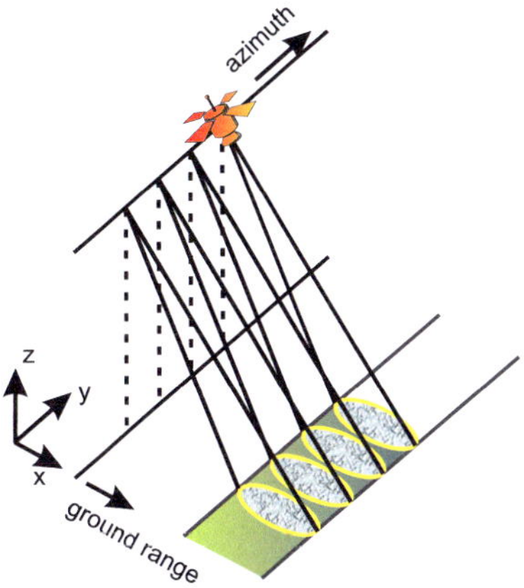

(a) strip-map mode

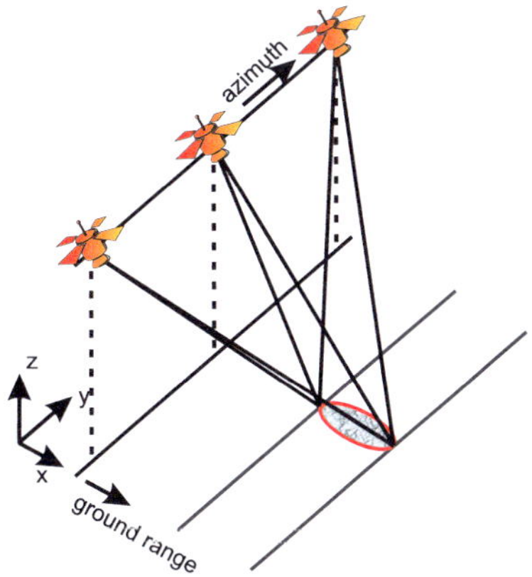

(b) spotlight mode

Figure 2.5: Geometry of a SAR system.

same ground spot over the whole illumination time. This allows an improvement of azimuth resolution in the resulting radar image. Fig. 2.5(b) shows the geometry of the spotlight mode. The length of the synthetic aperture, and by this the azimuth resolution, increases for all targets in the illuminated area through adequate steering of the antenna beam. The increase of the synthetic aperture is the same for all illuminated scatterers. The disadvantage of the spotlight mode is the limitation of the swath length to the projection of the antenna beam on the ground – a single spot. Continuous illumination of the ground swath is not possible in this mode. The improvement of the azimuth resolution is traded off by a loss of coverage in azimuth direction.

2.4.3 ScanSAR Mode

The scanSAR mode is optimized to improve the width of the illuminated swath in range. The whole swath is divided into subswaths as indicated in Fig. 2.6(a). Subswaths are illuminated with so called 'subbeams'. Receive data of all 'subbeams' are combined in order to achieve a continuous image in signal processing. For every subswath the radar sends a group of chirps called 'burst', followed by a gap. In this gap data from other subswaths are collected. In the scanSAR mode only a part of the targets' Doppler history

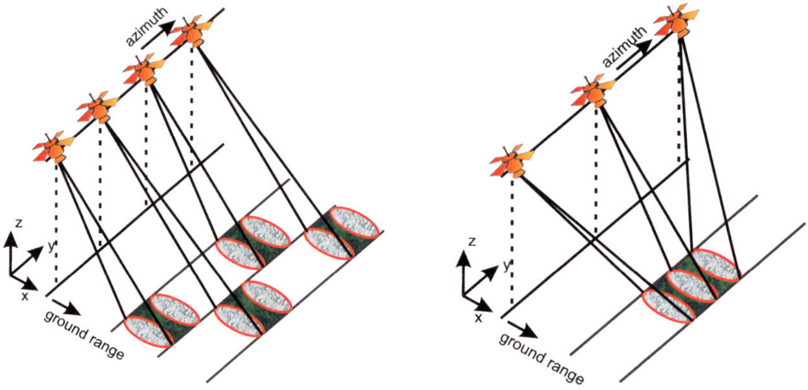

(a) scanSAR mode (b) sliding spotlight mode

Figure 2.6: Geometry of a SAR system.

is collected. As a result the azimuth resolution degrades. The degradation is proportional to the number of subswaths as the illumination time must be divided between all 'subbeams'. In the scanSAR mode a very wide swath in range is possible. Typically for spaceborne scanSAR mode each subswath covers ca. 100 km to 150 km, with a whole swath width of the scanSAR mode of up to 700 km, [12]. The swath width is limited by the range resolution at near range and the transmit power at far range. This mode is used in satellite as well as in airborne radars. In the scanSAR mode the azimuth resolution is traded off for wide swath width.

2.4.4 Sliding Spotlight Mode (Hybrid Strip-Map/Spotlight Mode)

The sliding spotlight mode, also called hybrid strip-map/spotlight mode, is a compromise between the high resolution in azimuth of the spotlight mode and the large coverage in azimuth direction of the strip-map mode. The spotlight mode improves the azimuth resolution of the radar image but the disadvantage is the non-continuous and limited coverage of the ground swath in azimuth direction. The strip-map mode enables continuous coverage but the azimuth resolution is limited to half the effective antenna length. A compromise is a sliding spotlight mode where the azimuth resolution is better than in the strip-map mode and at the same time the swath length is larger than in the spotlight mode. The radar antenna in the sliding spotlight mode is steered in azimuth at a point slightly farther away from the radar than the area being illuminated. The geometry of the sliding spotlight mode can be seen in Fig. 2.6(b). The strip-map and the spotlight mode can be considered as two limiting cases of the sliding spotlight mode. If the antenna is steered about the center of the illuminated scene the spotlight mode is achieved. The other limiting case is if the antenna is steered at a point in infinity for the strip-map mode. Through adequate steering of the radar antenna the illuminated swath length is larger but still allowing to improve the azimuth resolution with longer Doppler history of scatterers in the illuminated area. A compromise between the swath length and the azimuth resolution can be achieved. A more detailed description of this mode is given in Chapter 6 and Chapter 7.

2.5 Digital Beam Forming (DBF) in SAR

Another way to improve the performance of the SAR system is the use of a digital beam forming (DBF), [31], [46], [100]. The High-Resolution Wide-Swath (HRWS) SAR, described in this work, uses digital beam forming to improve performance by resolving the contradiction between high azimuth resolution and wide swath width, [14], [83], [84], [85], [102]. The HRWS SAR has a single transmit antenna and a receive antenna array consisting of M panels (in azimuth) each with N sub-apertures (in elevation). Fig. 2.7 illustrates the transmit and receive antenna system of the HRWS SAR.

Demands on radar performance are high resolution and wide coverage area. The range resolution depends on the used bandwidth of the transmitted signal. The azimuth resolution depends on the effective antenna length. To improve the azimuth resolution the antenna length has to be shorter. Improving the swath width requires on the other hand a smaller height of the radar antenna. In order to sample the radar receive data without ambiguities the height and the length of the radar antenna must fulfill the requirement for the minimum antenna aperture size A_p. The dependence between parameters for a single radar antenna is given in Eq. (2.19), [45]

$$A_p = l_{t_x r_x} \cdot h_{t_x r_x} > \frac{4 v_p \lambda r_s \tan(\varphi_i)}{c_0} \tag{2.19}$$

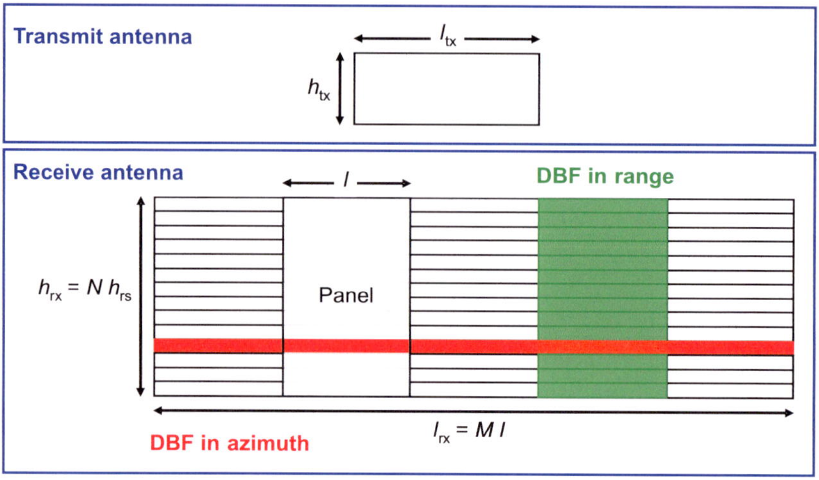

Figure 2.7: Configuration of the transmit and receive antenna array for the HRWS SAR.

Where $l_{t_x r_x}$ and $h_{t_x r_x}$ are the length and the height of the radar antenna, λ is the wavelength, v_p is the speed of the radar platform, c_0 is the speed of light, r_s is the slant range calculated to the middle of the illuminated swath and φ_i is the incidence angle under which the scene is illuminated. It is clear from Eq. (2.19) that with a constant antenna aperture size A_p a decrease of the length of the radar antenna must result in an increase of the height and vice versa.

The HRWS SAR has a separate transmit and receive antenna. This relaxes requirements for antennas as they can be optimized separately for transmit or for receive. The transmit antenna is a single aperture. The height of the transmit antenna is reduced compared to a mono-static SAR to obtain a wider swath on the illuminated ground. The lower single antenna gain can be compensated by large antenna aperture of the whole array and high efficiency of the front-end. The lower PRF resulting from DBF in azimuth direction ensures unambiguous sampling of the illuminated scene. Some basic limitations for antenna design issues are presented in [99]. The receive antenna consists of multiple sub-apertures each with independent receive front-end. A possible modification is to group some of the receive sub-antennas and provide a single receive front-end, which makes the antenna system less complicated and reduces costs.

The HRWS SAR uses digital beam forming in both elevation and azimuth direction. Fig. 2.8 presents a general diagram of DBF processing. First the DBF in range is applied for each panel of the receive antenna array. Signals from all sub-apertures from one panel are coherently added to create a narrow beam which is swept over the illuminated ground. The algorithm is called scan-on-receive. A detailed description of this algorithm follows in Section 2.5.1. The DBF in azimuth is applied on the receive data after the DBF in range, as indicated in Fig. 2.8. Two algorithms that can be used for DBF in azimuth direction are described in Section 2.5.2.

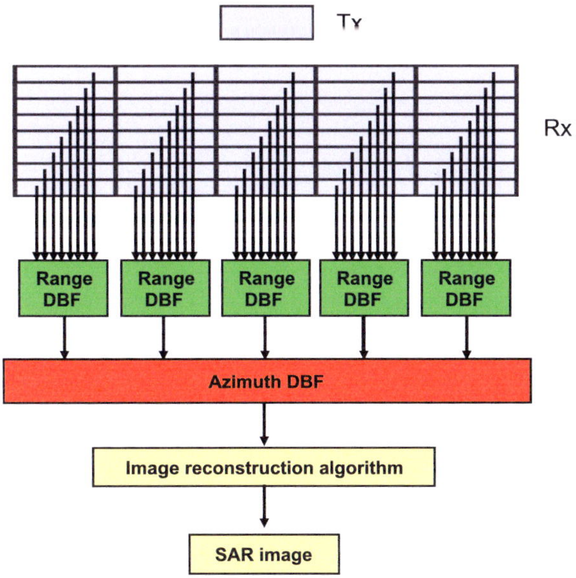

Figure 2.8: General diagram of the DBF processing in azimuth and range direction.

2.5.1 Range DBF

The HRWS SAR receive antenna has N sub-apertures with independent receive channels. This configuration, with an adequate DBF algorithm, allows to create a narrow beam in elevation and slide it over the ground or to point it at a specified direction. The HRWS SAR uses a DBF algorithm in elevation called scan-on-receive, [31], where the receive beam slides over the ground as shown schematically in Fig. 2.9. A single array element has a broad beam in elevation to cover the whole swath in range. Through DBF a narrow beam is created which slides over the ground.

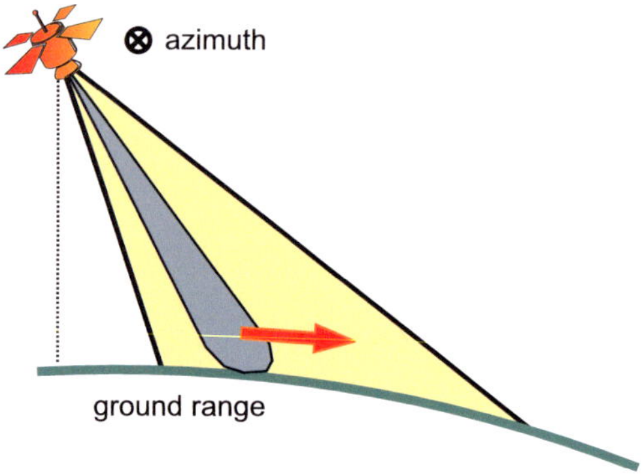

Figure 2.9: The digital beam forming scan-on-receive in range.

The scan-on-receive algorithm consists of two phase shifts, each realized as multiplication with an exponential function and following summation of all receive signals. The first multiplication takes place in the time domain and the phase of the exponential function is time dependent. The second multiplication can be applied in the frequency domain with a frequency dependent phase. The second multiplication is necessary only for radar signals with very wide bandwidth. The error made by not applying the second multiplication for a moderate bandwidth radar signal is negligible.
The time dependent phase shift for each sub-antenna signal can be calculated according to Eq. (2.20).

$$\gamma_n(t_{rg}) = \frac{2\pi}{\lambda} \cdot d_n \sin(\theta(t_{rg})) \qquad (2.20)$$

d_n is the position of the nth sub-aperture in the array in meters and the angle $\theta(t_{rg})$ can be calculated according to Eq. (2.21). It is to be noted that the equation includes also the Earth's curvature. For spaceborne systems with high resolution the ground profile should be included in the phase estimation. Assuming the geometry as shown in Fig. 2.9 the θ can be calculated according to Eq. (2.21). The radar processing should include the ground elevation model (DEM - digital elevation model) for illuminated surfaces with significant height changes as e.g. mountains.

$$\theta(t_{rg}) = \arccos\left(\frac{4R_{Orbit}^2 - 4R_{Earth}^2 + t_{rg}^2 c_0^2}{4R_{Orbit} t_{rg} c_0} - \beta\right) \tag{2.21}$$

t_{rg} is the time in range with the start time at the beginning of the signal transmission, R_{Earth} is the medium Earth's radius, R_{Orbit} is the radius of the radar platform orbit. β is the NaDir angle which can be calculated according to Eq. (2.22).

$$\beta = \text{asin}\left(\frac{R_{Earth}\sin(\varphi_i)}{R_{Orbit}}\right) \tag{2.22}$$

The phase shift is applied to each signal received through each sub-aperture $s_n(t_{rg})$ as in Eq. (2.23).

$$s_n'(t_{rg}) = s_n(t_{rg})\exp(j\gamma_n(t_{rg})) \tag{2.23}$$

The second multiplication is applied in the frequency domain. The phase shift can be calculated according to Eq. (2.24) for each of the sub-aperture receive signals. The multiplication compensates effects caused by the bandwidth extension of the radar signal. The time dependent phase in Eq. (2.20) is estimated for a center frequency of the radar signal.

$$\gamma_{fn}(f_{rg}) = \frac{2\pi}{\lambda} d_n \sin\left(\tau \frac{f_{rg}}{B}\theta_f\right) \tag{2.24}$$

where f_{rg} is the range frequency, B is the bandwidth of the transmitted chirp, θ_f can be calculated according to Eq. (2.25).

$$\theta_f = \frac{-0.5 \cdot c_0(b_2 - b_1 - R_{Orbit}^2 + R_{Earth}^2)}{R_{Orbit}^2 \sin(\beta)(2b_2 - b_1 - R_{Orbit}^2 + R_{Earth}^2)} \tag{2.25}$$

where b_1 and b_2 are defined as in Eq. (2.26) and in Eq. (2.27) accordingly.

$$b_1 = R_{Orbit}\cos(\beta)\sqrt{R_{Orbit}^2 \cos^2(\beta) - R_{Orbit}^2 + R_{Earth}^2} \tag{2.26}$$

$$b_2 = R_{Orbit}^2 \cos^2(\beta) \tag{2.27}$$

Radar signals of each sub-aperture are multiplied with the second phase in the frequency domain calculated according to Eq. (2.24). The radar receive data is first transformed into the frequency domain through Fourier transformation. Then the radar signal for each sub-aperture is multiplied with the adequate exponential function as in Eq. (2.28).

$$S_n''(f_{rg}) = S_n'(f_{rg}) \exp(j\gamma_{fn}(f_{rg})) \tag{2.28}$$

Radar receive signals are superpositioned over all sub-apertures in elevation as in Eq. (2.29). If the second phase multiplication has been applied to the radar signal the summation can also be performed in the frequency domain as the Fourier transformation is a linear operation. The transformation of received signals into the frequency domain must be included in the following SAR processing.

$$s(t_{rg}) = \sum_{n=1}^{N} s_n''(t_{rg}) \tag{2.29}$$

The DBF in elevation can be applied after the radar data is transmitted to the ground station. Then the DBF algorithm can be easily adjusted to include also, e.g., the ground profile for highly resolved radar data. The DBF in range can also be performed directly on board of a satellite. The advantage of this solution is a significant reduction of the data amount which has to be transmitted to the ground station. The HRWS SAR receives an independent radar signal for each sub-aperture. With an array of N elements in elevation this makes N times the data amount as for a single receive antenna. After DBF processing in elevation the amount of data is reduced to the same amount as for a radar with a single aperture in elevation.

2.5.2 Azimuth DBF

The DBF algorithm in azimuth direction is applied on receive signals in the HRWS SAR after the DBF processing in range. The range DBF processing combines receive signals from all sub-apertures in elevation reducing by this the amount of radar data. In azimuth DBF the amount of radar data is usually not changed or the change is not significant. The idea for the azimuth DBF is to sample the radar receive signal in azimuth direction multiple times with sub-apertures aligned in the flight direction of the radar platform at the same

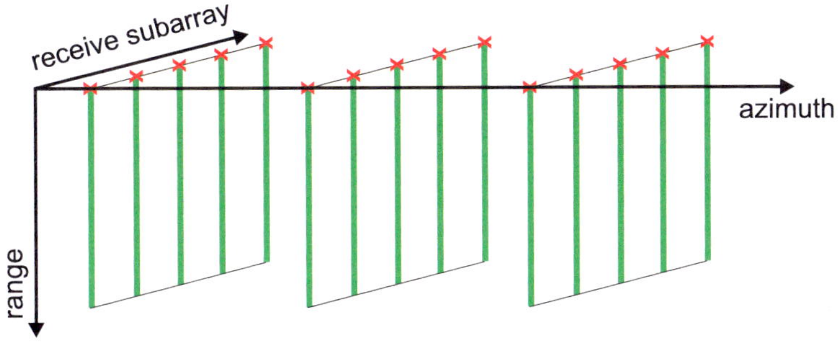

Figure 2.10: Graphical illustration of the 3 dimensional data space of SAR raw data with multiple independent sub-apertures in azimuth.

time [24], [29]. The requirement for the minimal PRF can then be relaxed. The sampling theorem for correct unambiguous sampling demands a specified number of sampling points in a specified time interval but those samples do not have to be uniformly distributed over this interval [17], [67], [72]. The SAR signal can be reconstructed from a non-uniform sampling of an illuminated scenario [27], [26]. Collecting radar receive data for each azimuth sub-aperture gives M azimuth samples for each position of the radar platform. Conventional SAR raw data can be written into a two-dimensional data matrix with one dimension being the range and the other one being the azimuth direction. The raw data collected by a radar system with DBF in azimuth can be seen as a three-dimensional matrix. The dimensions are: range, azimuth and the number of receive sub-antenna in azimuth as shown in Fig. 2.10. In general the DBF algorithm in azimuth can be seen as mapping the three-dimensional data matrix into a two-dimensional maintaining all advantages of a multi-aperture radar system. The sampled radar data can be reconstructed with an adequate azimuth DBF algorithm and processed with a strip-map SAR algorithm, [28]. Experimental investigation of the performance of various DBF algorithms can be found in [38], [39], [40]. The two azimuth DBF algorithms are described in the following.

The first algorithm for azimuth DBF is based on properties of the Fourier transformation. A detailed description of this algorithm can be found in [97]. The main steps in this algorithm are a phase shift to correct the phase difference between a bi-static and a mono-static SAR configuration, correction for a non-uniform azimuth sampling and an appropriate concatenation of the receive data, which maps the three-dimensional data space into a two-dimensional one. After this step the raw radar data can be compressed with an algorithm for a mono-static strip-map SAR. A schematic diagram of this azimuth DBF algorithm is presented in Fig. 2.11(a).

Fig. 2.11(b) shows the diagram of the second azimuth DBF algorithm. A detailed description of this DBF algorithm can be found in [24], [25], [43]. The idea behind this algorithm is that each receive signal from a sub-antenna is a subsampled original signal. The spectrum of each of these signals is shifted by ±PRF spectra of the same signal. Through an adequately designed filter bank the spectrum of the original signal can be reconstructed.

An example with two sub-apertures in azimuth with design equations for the filter bank for both of the receive antenna signals is described in following. The filters $H_1(f_{az})$ and $H_2(f_{az})$ transform the original radar signal of a strip-map SAR to the receive signal of a DBF SAR with two radar sub-antennas in azimuth. As the number of samples per separate antenna is lower for the DBF SAR, only an adequate number of samples after those filters is considered. The d_1 and d_2 are positions of sub-apertures calculated from the middle of the array. The filters $H_1(f_{az})$ and $H_2(f_{az})$ can be defined as in Eq. (2.30) and Eq. (2.31) accordingly.

$$H_1(f_{az}) = \exp\left(-j\pi\left(\frac{d_1^2}{2\lambda r_0} + \frac{d_1}{v_p} f_{az}\right)\right) \tag{2.30}$$

and

$$H_2(f_{az}) = \exp\left(-j\pi\left(\frac{d_2^2}{2\lambda r_0} + \frac{d_2}{v_p} f_{az}\right)\right) \tag{2.31}$$

Writing these equations into a matrix and including the shift in spectrum the matrix $H(f_{az})$ is obtained.

$$H(f_{az}) = \begin{bmatrix} H_1(f_{az}) & H_2(f_{az}) \\ H_1(f_{az} + \mathrm{PRF}) & H_2(f_{az} + \mathrm{PRF}) \end{bmatrix} \tag{2.32}$$

To reconstruct the original mono-static strip-map SAR receive signal the matrix from Eq. (2.32) is inverted.

$$\begin{aligned} A_H &= H_1(f_{az}) \cdot H_2(f_{az}) \cdot \ldots \\ &\left(\exp(-j\pi\frac{\Delta y_2}{v_p}\mathrm{PRF}) - \exp(-j\pi\frac{\Delta y_1}{v_p}\mathrm{PRF})\right) \end{aligned} \tag{2.33}$$

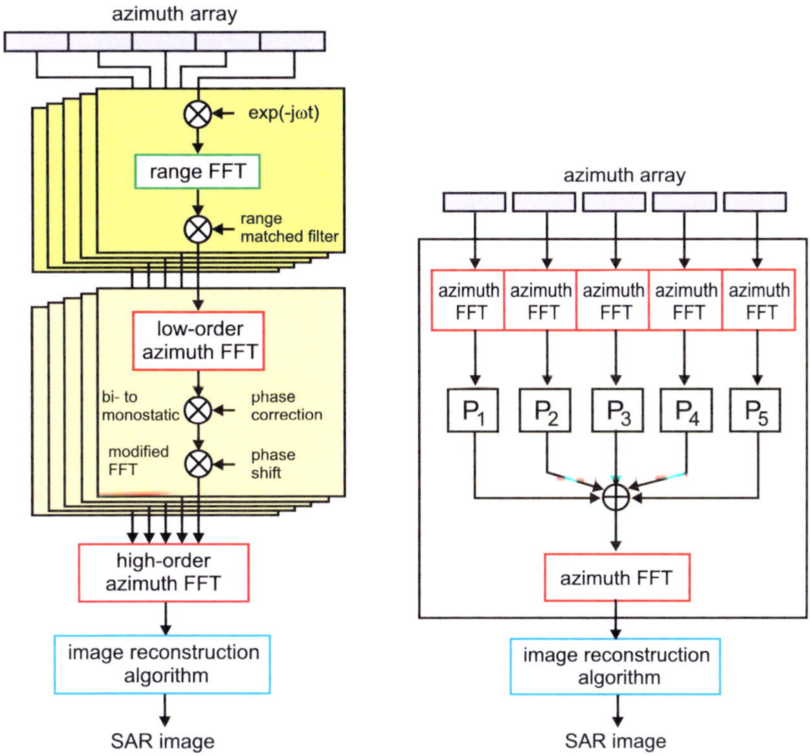

(a) algorithm based on the Fourier transformation

(b) algorithm based on filter bank

Figure 2.11: General block diagrams for DBF in azimuth.

$$H(f_{az})^{-1} = \frac{1}{A_H} \begin{bmatrix} H_2(f_{az} + \mathrm{PRF}) & -H_2(f_{az}) \\ -H_1(f_{az} + \mathrm{PRF}) & H_1(f_{az}) \end{bmatrix} \tag{2.34}$$

The filter bank is defined from the inverse $H(f_{az})$ matrix as in Eq. (2.35) and Eq. (2.36).

$$H(f_{az})^{-1} = P = \begin{bmatrix} P_{11} & P_{12} \\ P_{21} & P_{22} \end{bmatrix} \tag{2.35}$$

$$\begin{aligned} P_1 &= [P_{11} P_{12}] \\ P_2 &= [P_{21} P_{22}] \end{aligned} \tag{2.36}$$

With filters P_1 and P_2 the original receive signal can be reconstructed. The block diagram of the reconstruction algorithms is presented in Fig. 2.11(b). The reconstructed receive radar data can then be processed with a strip-map SAR processing algorithm.

2.6 Ideal Point Target

A SAR system can be considered a linear system. The simplest way to characterize a linear system is its impulse response, [12], [58], [73]. The quality of the SAR system and of the signal processing can be determined by a measurement or a simulation of a point target. In SAR the impulse response of the system is obtained by measuring a response from a single, isolated strong scatterer on the ground called a point target or a point scatterer. The performance of the SAR system should be characterized not only in the middle of the illuminated swath but also on edges of the swath in range and for modes other than strip-map also in azimuth direction, as the radar system can have slightly different performance depending on the position of the point target (range, azimuth and height). A very good approximation of a point target for measurement purposes is a corner reflector.

An ideal point target in a SAR image is described with a two-dimensional $\sin(x)/x$ function. It is to be noted that this is the ideal case with infinite time and bandwidth extension. The receive signal for a point target measurement or simulation is limited in time and in frequency. Fig. 2.12(a) shows the two-dimensional representation of an point target and Fig. 2.12(b) shows the one-dimensional cut in range. It is to be noted that the response is symmetrical in both range and azimuth direction. Fig. 2.13(a) and Fig. 2.13(b) show the magnitude and the phase of the two-dimensional Fourier transformation of the point target accordingly. The spectrum of the point target is concentrated around frequencies $|f_{rg}| < B_{rg}$ and $|f_{az}| < B_{az}$, where B_{rg} and B_{az} are the range and azimuth bandwidth of the received radar data accordingly. The magnitude of the spectrum of the point target is constant as it can be seen in Fig. 2.13(a). The phase of the point target's spectrum presented in Fig. 2.13(b) is linear in both range and azimuth frequency. If the target is positioned exactly in the middle of the transformed area the phase characteristic is constant. If the point target is shifted slightly from the middle the phase characteristic is linear, and the slope depends on the shift length from

the middle of the transformed scene. This effect can be seen in Fig. 2.14. The Fig. 2.14(a) shows the magnitude of the point target positioned slightly shifted of the middle of the transformed area in both range and azimuth. Fig. 2.14(b) shows the phase characteristic of this point target. The phase changes linearly in both range and azimuth frequency resulting in the stripe pattern.
The impulse response gives the opportunity to quantify the SAR system and the SAR processing algorithm performance. The most important figures of merit for a SAR system are range and azimuth resolution, peak-to-side lobe ratio (PSLR), integrated side lobe ratio (ISLR), peak amplitude, SNR and noise floor. A short description of each figure of merit can be found in following sections.

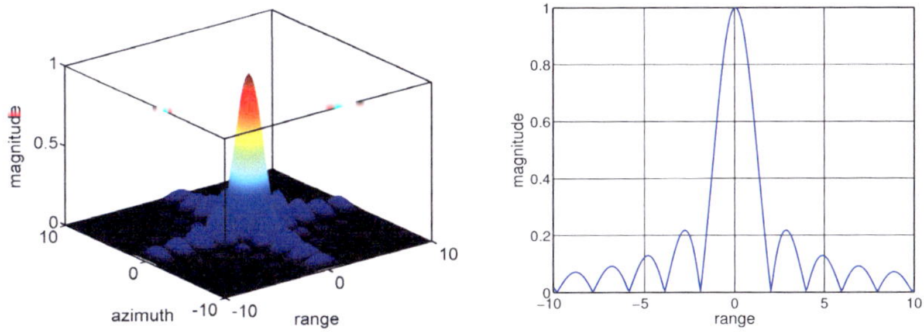

(a) two dimensional representation

(b) cut in range direction

Figure 2.12: Normalized magnitude of an ideal point target.

2.6.1 Resolution - 3dB Beam Width

One of the most often used figures of merit for a radar system is resolution. The resolution of a radar image is defined as the 3 dB impulse response width of the compressed point target. Fig. 2.15 shows a cut in range direction through an ideal two-dimensional point target response. The 3 dB width of the main lobe in range direction is marked with a horizontal arrow. The azimuth resolution is defined as the 3 dB width of the main lobe in azimuth direction. The range resolution depends on the bandwidth of the transmitted radar signal. The azimuth resolution for SAR systems depends on the effective length of the antenna. Both equations were already described in

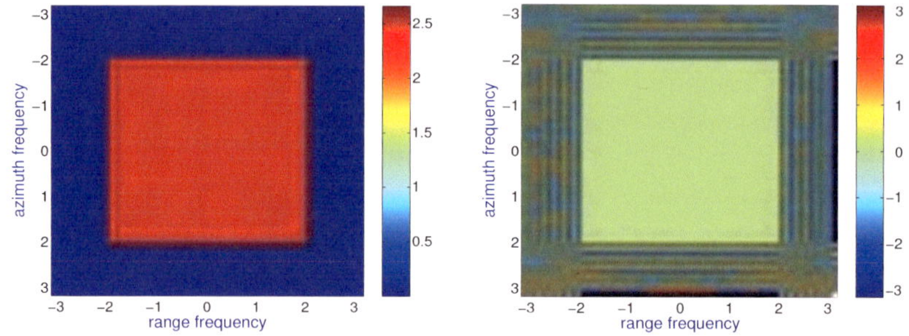

(a) magnitude of the spectrum (b) phase characteristic

Figure 2.13: Two dimensional Fourier transform of an ideal point target.

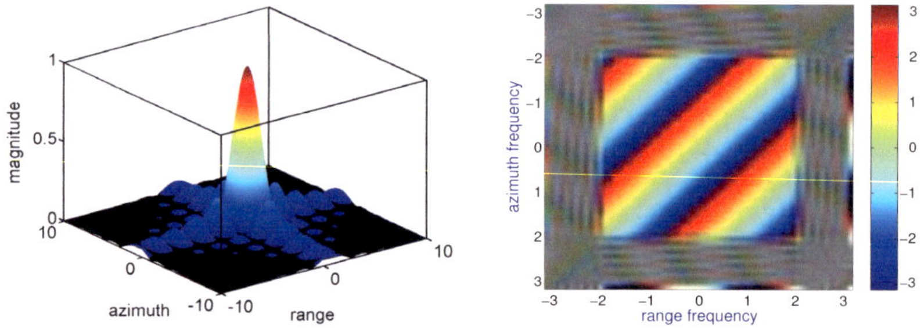

(a) magnitude of a point target (b) phase characteristic

Figure 2.14: An ideal point target shifted in range and in azimuth direction.

Section 2.2.3. Eq. (2.13) gives the resolution in range and Eq. (2.18) gives the resolution in azimuth for a SAR system working in the strip-map mode. A SAR image is often weighted with a window function in both range and azimuth direction. This reduces the height of side lobes. The price for application of the windowing is a widening of the main lobe. The resolution in range and in azimuth with applied tapering window has to be multiplied with an adequate scaling factor. This factor depends on the used window, which for a rectangular window is equal to one. The factor for all common windows used in radar systems can be found in literature, [12]. In this work no weighting is used for the radar signal (rectangular window).

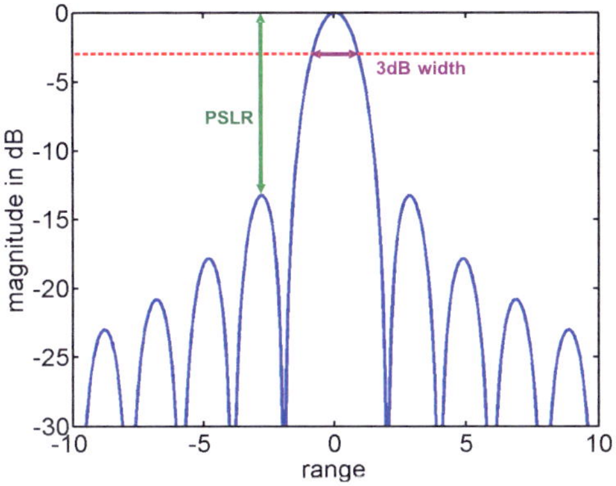

Figure 2.15: Point target with marked 3 dB impulse response width and PSLR

2.6.2 Peak-to-Side Lobe Ratio

Another important figure of merit for a point target is the peak-to-side lobe ratio (PSLR). The PSLR is defined as the ratio between the main lobe and the highest side lobe. The ratio is usually given in decibels. In Fig. 2.15 the PSLR is marked with a vertical arrow. For an ideal point target, which appears as a sinc-function, the PSLR is about 13.4 dB. This can be improved with the help of a tapering window, for example Hamming, in the signal processing chain. Through this procedure of appropriate windowing of the radar signal also small targets, which would be masked otherwise by an adjacent staring targets, are visible in the SAR image.

2.6.3 Integrated Side Lobe Ratio

The integrated side lobe ratio (ISLR) is defined as the integrated power of the side lobes over the integrated power of the main lobe. The one-dimensional ISLR can be calculated according to Eq. (2.37).

$$ISLR = 10 \log_{10} \left(\frac{P_{tot} - P_{main}}{P_{main}} \right) \tag{2.37}$$

P_{tot} is the total power and P_{main} is the 'main-lobe' power. P_{main} is equal to an integral over the main lobe. The limits for the integration can be defined in

various ways. The most common approach is to integrate over the 3 dB main lobe width (see 2.6.1) multiplied by a constant parameter α. The constant α is usually chosen between 2 and 2.5. Sometimes a null-to-null definition is used in order to calculate the main lobe power.
The two-dimensional ISLR is defined similarly to the one-dimensional one. The main lobe is defined within a rectangle of the length α times the main lobe 3 dB width in each direction. The side lobe power is then the power of the entire two-dimensional image minus the main lobe power.

2.6.4 Peak Amplitude

The signal peak amplitude is the maximal amplitude of the impulse response. It is an important figure of merit used later for comparing various simulations. If the compression is not fully correct the peak energy 'leaks' into the neighboring cells and the maximum of the peak amplitude decreases. The peak amplitude can be used, e.g., to determine the influence of a mechanical distortion on an antenna array plane. In following chapters the drop of the peak amplitude is calculated in comparison to a reference simulation.

2.6.5 Signal-to-Noise Ratio

The signal-to-noise ratio (SNR) is a measure of signal's quality used broadly in engineering and science. The SNR describes the power of the desired signal compared to the power of the noise. Eq. (2.38) describes the general formula for the SNR. P_{signal} is the power of the wanted signal, P_{noise} is the noise power.

$$SNR = \frac{P_{signal}}{P_{noise}} \tag{2.38}$$

In SAR the increase in the number of receive samples improves the SNR of the image, [97]. This can be considered as a part of processing gain. The SNR_{image} of a single image resolution cell depends on the number of signal samples that are contributing to the processing of the cell. Resulting from the coherent SAR processing the SNR of a single resolution cell increases by n_{pulse}, which is the number of samples per chirp (fast time), and n_{az}, which is number of samples in the synthetic aperture (slow time). For systems with multiple independent receive apertures N the SNR increases in the ideal case by the number of independent receive sub-apertures. This relationship is expressed by Eq. (2.39).

$$\begin{aligned} SNR_{image} &= SNR_{signal} \cdot N_{processing} \\ &= SNR_{signal} \cdot n_{pulse} \cdot n_{az} \cdot N \end{aligned} \tag{2.39}$$

n_{pulse} depends on the duration of the chirp τ and the sampling rate f_s.

$$n_{pulse} = \tau f_s \tag{2.40}$$

n_{az} depends on the length of the radar synthetic aperture X, the sampling frequency in azimuth PRF and the radar platform velocity v_p.

$$n_{az} = \frac{X}{\Delta u} = \frac{X}{v_p/\text{PRF}} \tag{2.41}$$

Δu is the azimuth sample spacing. The SNR is usually expressed in decibels.

2.6.6 Average Noise Floor Level

The average noise floor level is a measure for quality of the radar image. It gives the ratio of the main maximum in the radar image caused by a target to the mean noise level [82]. The average noise floor level is usually given in dB. This parameter takes into account radar image noise resulting from input noise, receiver noise including quantization noise and processing gain. The side lobes of the reflecting target are not accounted in the calculation of the average noise floor level. In this work the average noise floor level is used to measure the level of unwanted signal or distortion after the SAR processing.

3 Influence of Radar Receiver on SAR Image

The transmitter and the receiver front-end are important parts of radar and communication systems. The trend nowadays in communication and radar technology is to shift the digitalization of the transmit and the receive signal as far as possible forward to the transmit/receive antenna. This makes the front-end simpler and more flexible. Still some analog components are necessary especially for higher frequencies. The analog part of the radar receiver and its influence on the received radar signal and the compressed image, the error sensibility of the system are considered in this chapter. A direct conversion receiver is taken as a model. The receive front-end consists of bandpass and lowpass filter bank, amplifiers, an IQ-mixer and an analog-to-digital converter (ADC). A new way to describe receiver components in the frequency domain is introduced called frequency domain VKA-representation. This representation provides an efficient and simple way to describe electronic components including nonidealities and coupling between I and Q path of the receiver. The representation is based on a similar description used for an OFDM (Orthogonal Frequency Division Multiplex) telecommunication system [48].

The direct conversion receiver and its components are described in Section 3.1. The VKA-representation in time domain and in frequency domain is presented in Section 3.2. All components of the radar receiver are described in the proposed frequency domain VKA-representation in Section 3.3. In the following Section 3.4 the general model of nonidealities in the radar front-end is given. The HRWS SAR demonstrator provided by Astrium GmbH was measured to improve components models. Measurements were recalculated to the new VKA-representation. Results of the measurement are described in Section 3.5. A simulation has been developed for a SAR system including receiver front-end element models. This simulation is described shortly in Section 3.6. Finally results of simulations with selected example nonidealities are presented in Section 3.7.

3.1 Receiver Front-End: Direct Conversion Receiver

The receiver of a radar system is modeled on the basis of a direct conversion receiver [91]. This type of receiver is usually used in radar systems. It was also used in the HRWS SAR demonstrator built by Astrium GmbH to show the performance of a multi-channel low bit radar receiver [32]. The HRWS SAR demonstrator can be seen in Fig. 3.1. The demonstrator consists of eight identical independent receive channels. Measurements of the demonstrator components are presented in the following Section 3.5. The direct conversion receiver, as the name indicates, is down-converting the input signal directly into baseband without any intermediate stages. The direct conversion of frequency reduces the complexity of hardware, but puts more demands on the performance and the quality of transmitter and receiver components. In the transmitter the radar signal is amplified, filtered and up-converted to the required transmit frequency band. Then the radar signal is again filtered and amplified before transmitting with the transmit antenna. The transmitted radar signal is reflected on a target. The reflected signal is then received with the receive antenna. The radar receiver consists of the same or similar components, where the radar received signal is going though all components in reverse order. The block diagram of the direct down-conversion receiver is presented in Fig. 3.2. The received signal is amplified with a low noise amplifier (LNA) and bandpass filtered. The receive signal is down-converted into the baseband with an IQ-mixer. Signals in both channels, I and Q, are again filtered and amplified before digitalization with an ADC. Further processing of the signal is done through a digital signal processing.

3.1.1 Components of the Receiver

In following presented is a short description for components of a direct conversion receiver as shown in Fig. 3.2: amplifiers, bandpass and lowpass filters, a local oscillator (LO), an IQ-mixer and an ADC.

Amplifiers

More than one amplifier type is usually present in a receiver chain. In Fig. 3.2 for both I and Q paths the receive signal is going through three stages of am-

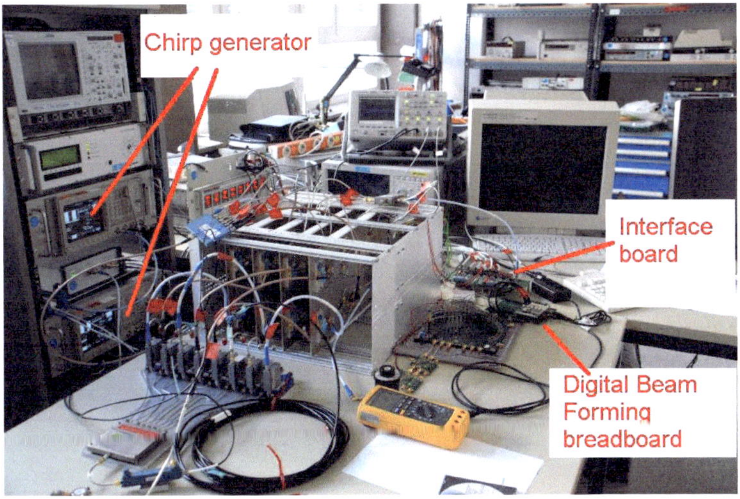

Figure 3.1: HRWS SAR demonstrator

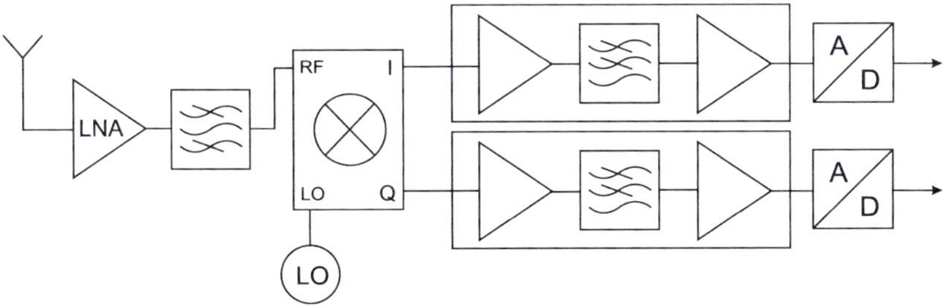

Figure 3.2: Block diagram of a direct down conversion receiver as implemented in the HRWS SAR demonstrator.

plification. A low noise amplifier (LNA) is located directly before or after receive antenna, see Fig. 3.2. In this type of amplifier, as the name indicates, the noise figure is optimized to be very low. The LNA is sensitive to overvoltage. In order to protect the amplifier a limiter is placed usually just before it in the receiver chain. The LNA in the HRWS SAR receiver works in X-band, at a frequency of about 9.65 GHz. Amplifiers, which are placed after the IQ-mixer in the receiver, work in the baseband. They are optimized for high amplification of the input signal. In Fig. 3.2 two stages of amplifiers in the baseband are visible, which correspond to the design of the HRWS SAR demonstrator receiver. Baseband amplifiers ensure optimal steering of the following ADC.

Filters

Another important element of the receiver chain are filters. A typical transmitter or receiver includes a lowpass filter and a bandpass filter. A short description of both filters follows.

Bandpass filter The bandpass filter is used to cut out wanted part of the received signal. The bandpass filter is usually placed after the receive antenna to limit the frequency spectrum of the signal. A radar system transmits a signal with a specific bandwidth and carrier frequency. The Doppler shifts caused by moving targets are relatively small for a SAR system compared to the bandwidth of the radar signal. The bandpass filter cuts from the received radar signal all frequencies lower or higher than the wanted signal bandwidth. The middle frequency of the bandpass filter is usually the carrier frequency of the transmitted signal. This allows to avoid possible unwanted mixing products in demodulation with frequencies other than the wanted frequency band.

Lowpass Filter Lowpass filters cut off all frequencies above a defined cut-off frequency and let pass only frequencies under this threshold. Those filters are placed after quadrature down–conversion mixer when the receive signal is in the baseband. The filter removes higher order harmonic components which are present in the signal after mixing of the receive signal with a local oscillator (LO) signal. The signal resulting after the filtering step is given at the input of the ADC. The filter ensures that signals at the input of the

ADC do not contain components with frequencies higher than half of the sampling frequency. Otherwise the high frequency components of the input signal would disturb wanted signal in the digital domain (aliasing effect).

Local Oscillator

The local oscillator (LO) generates a signal at a specified frequency. In the considered system the LO generates a sinusoidal signal with a constant frequency of $f_c = 9.65$ GHz in X-band.

IQ-Mixer

The IQ-mixer also called quadrature demodulator down–converts a high frequency input signal into inphase and quadrature components in baseband. A simplified diagram of the IQ-demodulator can be seen in Fig. 3.3. The IQ-demodulator can be represented by a system consisting of two mixers, a 90° phase shifter and a local oscillator (LO). The input signal is mixed once with the LO signal (I-component) and once with the 90° shifted LO signal (Q-component). In the direct down conversion receiver the I and Q components are down converted directly into the baseband. The I and Q components are a complex valued signal representation. The resulting down-converted signal can be described with Eq. (3.1). s_i is the inphase component and s_q is the quadrature component.

$$s = s_i + js_q \tag{3.1}$$

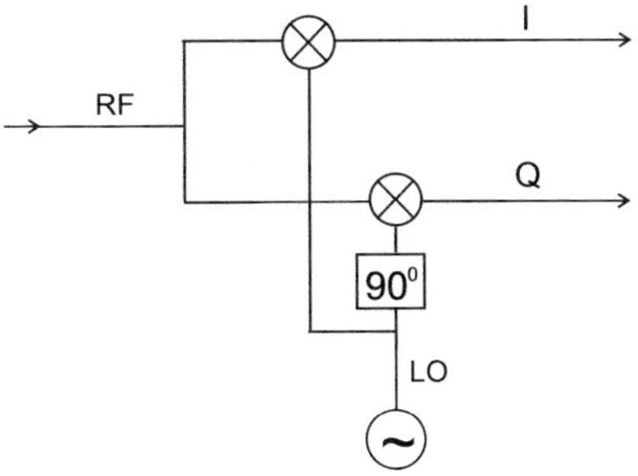

Figure 3.3: Block diagram of a quadrature demodulator (IQ mixer).

Analog-To-Digital Converter

The analog-to-digital converter (ADC) converts an analog input signal into its digital representation. Both time and amplitude of the input signal are discretized. Important for proper quantization of a signal with the ADC is that the signal is band–limited with no frequencies higher than half the sampling frequency. The maximal amplitude of the input signal is on one hand limited, and does not exceed the maximal input amplitude, but on the other hand is high enough to optimally use the dynamic range of the ADC. A detailed description of quantization processes can be found in Chapter 5. Parameters of the ADC are the number of quantization bits b and the maximal amplitude A_m, above which the ADC goes into saturation.

3.2 Description of Receiver Components

A new representation of the receiver components is presented in this section. The representation consisting of three parameters is sufficient to describe all components of the receiver front-end. A general concern while designing simulations is how to describe electronic elements including their functionality and nonidealities in a simple and effective way. The representation should be able to describe various effects and distortions that occur in the receive front-end, but still be simple and easy to operate on. The new representation should describe effects of a single receiver channel as well as of multi-channel receiver effects between those channels. It is also important that it remains efficient in computing time for a simulation, while still describing accurately electronic components of the transmit and the receive front-end. It would be also of advantage to describe both I and Q paths of an IQ–receiver at the same time. The representation can be used throughout the whole transmit and receive front-end.
The VKA-representation in the time domain is used in [48] for the description and compensation of errors in an OFDM communication system. In a communication system including the transmitter and receiver front-end, both digital and analog parts, can be described with the help of the VKA-representation including the nonidealities of the system. Using the same representation an efficient method of error compensation for an OFDM system can be applied. In this work the VKA-representation is adapted to model elements of a radar receive front-end. It is simpler for a radar system to define the VKA-parameters

in frequency domain. For this a modified VKA-representation in the frequency domain is proposed. In Section 3.2.1 described is the VKA-representation in the time domain. In Section 3.2.2 a proposal for a new VKA-representation in the frequency domain is presented.

3.2.1 VKA-Representation in Time Domain

The VKA-representation in the time domain describes a single component or a block of components with the help of three parameters. These parameters are called in this work v_T, k_T and a_T. The index $_T$ indicates that the parameter describes the time domain model. The variable v_T is the weight for the original signal, variable k_T is the weight for the conjugate complex signal and a_T is an additional additive factor. The model operates on complex valued signals and can therefore describe with those three parameters both paths of quadrature modulated signals at the same time. In [48] a detailed description of all components of MIMO communication system representations can be found. Fig. 3.4 shows a block diagram of the time domain VKA-representation. $x(t)$ is the input signal, $y(t)$ is the output signal; g_R and g_I are multiplicative factors for the real and imaginary part of the signal (I and Q paths of the quadrature demodulated signal). In case of a real valued signal g_I is equal to zero. An electronic component can be represented with multiplications and additions as in Eq. (3.2).

$$\begin{aligned} y(t) &= g_R \cdot \Re\{x(t)\} + j g_I \cdot \Im\{x(t)\} + a_T \\ &= \frac{g_R + g_I}{2} x(t) + \frac{g_R - g_I}{2} x(t)^* + a_T \\ &= v_T x(t) + k_T x(t)^* + a_T \end{aligned} \tag{3.2}$$

Factors v_T, k_T and a_T can describe most receiver front-end components, and in particular types of effects in receiver e.g. IQ imbalance or coupling. These factors can be constant, time or frequency dependent. An important issue for a SAR system is the dependence of the systems components on frequency. In Fig. 3.4 all parameters are presented as frequency dependent. In the following section described is a modification of the VKA-representation for the frequency domain.

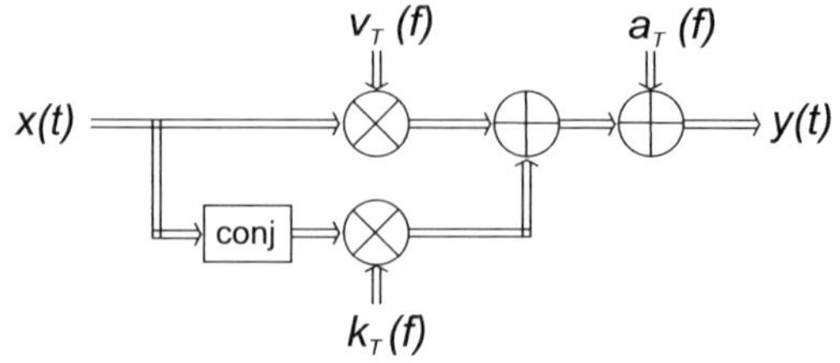

Figure 3.4: The VKA-representation in time domain.

3.2.2 VKA-Representation in Frequency Domain

Hardware components of a radar system usually have frequency dependent characteristics, which cannot be ignored. Some components, as e.g. amplifiers or filters, are more efficiently described in the frequency domain than in the time domain. The proposed VKA-representation in the frequency domain [63] is presented as a diagram in Fig. 3.5. Factors of the modified representation in frequency are called: V_F for the weighting of the original signal spectrum, K_F for the weighting of the reversed conjugate complex signal spectrum and A_F for an additive factor representing the additive component in the frequency domain. $X(f)$ is the spectrum of the input signal, $Y(f)$ is the spectrum of the output signal. The spectrum of the input and output signal can be calculated according to Eq. (3.3).

$$\begin{aligned} X(f) &= FFT\{x(t)\} \\ Y(f) &= FFT\{y(t)\} \end{aligned} \tag{3.3}$$

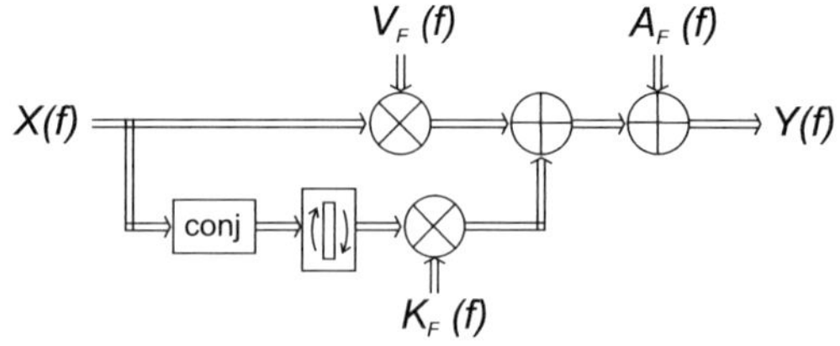

Figure 3.5: The VKA-representation in frequency domain.

The output signal $Y(f)$ can be calculated according to Eq. (3.4). G_R and G_I are characteristics in the frequency domain for the I and Q paths. Both characteristics are usually frequency dependent.

$$\begin{aligned} Y(f) &= G_R \cdot \Re\{X(f)\} + jG_I \cdot \Im\{X(f)\} + A_F \\ &= \frac{G_R + G_I}{2} X(f) + \frac{G_R - G_I}{2} X(-f)^* + A_F \\ &= V_F X(f) + K_F X(-f)^* + A_F \end{aligned} \tag{3.4}$$

I should be noted that the signal in the conjugated path is reversed before summation with the signal in other path.
The factors V_F and V_K of the frequency dependent VKA-representation can be calculated according to Eq. (3.5).

$$\begin{aligned} V_F &= \frac{G_R + G_I}{2} \\ K_F &= \frac{G_R - G_I}{2} \end{aligned} \tag{3.5}$$

The factors V_T, K_T, A_T and V_F, K_F, A_F are usually vectors or matrices dependent on time or frequency. It is possible to approximate those factors for a short time or narrow frequency span with a constant value.

3.3 Models of Receiver Components

Models for radar receiver front-end components are described in this section. The bandwidth in future high resolution SAR systems can reach the range of several GHz. The direct down-conversion receiver, which was described in Section 3.1, was used as a model. Models of amplifiers, filters and IQ mixer are presented in the frequency domain VKA-representation.

3.3.1 Amplifiers

Electronic components can be described with the help of S-parameters [78], [90], which describe the propagation of signals for all inputs and outputs of components. Here used is mostly the transmission parameter S_{12} describing the propagation from input (1) to the output (2). This S-parameter is referenced as S_{LNA} for the LNA. The LNA is positioned usually in the receiver right after the receive antenna before the IQ demodulation block, see Fig. 3.6. The VKA model of the LNA is described in Eq. (3.6). The $V_{F,LNA}$ is equal to the transmission parameter $S_{LNA}(f)$ for the LNA. $K_{F,LNA}$ is equal to zero as the LNA in before the IQ demodulation in the receive chain. $A_{F,LNA}$ is also equal to zero in ideal case.

$$\begin{aligned} V_{F,LNA} &= S_{LNA}(f) \\ K_{F,LNA} &= 0 \\ A_{F,LNA} &= 0 \end{aligned} \tag{3.6}$$

The VKA model for amplifiers positioned in the I and Q path, see orange marking in Fig. 3.6, is presented in Eq. (3.7). $V_{F,amp}$ represents the common part for both I and Q path. $K_{F,amp}$ describes the difference in transmission characteristics between both paths. $A_{F,IQamp}$ is equal to zero in ideal case.

$$\begin{aligned} V_{F,IQamp} &= \frac{S_{I,amp}(f) + S_{Q,amp}(f)}{2} \\ K_{F,IQamp} &= \frac{S_{I,amp}(f) - S_{Q,amp}(f)}{2} \\ A_{F,IQamp} &= 0 \end{aligned} \tag{3.7}$$

$S_{I,amp}$ and $S_{Q,amp}$ are the transmission S-parameters for the base-band amplifiers for the I and Q path accordingly.

3.3.2 Filters

Two different types of filtering is applied in the direct conversion receiver: one in the RF band directly at the input of the receiver with a bandpass filter, and a second one in the baseband after the IQ mixer with a lowpass filter. The bandpass filter suppresses all frequency components that are not in the band of the signal. The lowpass filter, which is applied after down conversion in the receiver, suppresses the higher harmonics of the down converted signal. Filters are characterized usually in the frequency domain with the amplitude and phase characteristic or in the time domain by their impulse response.
The bandpass filter VKA model is described in Eq. (3.8). $V_{F,BP}$ is equal to the transmission S-parameter of the filter $S_{BP}(f)$. $A_{BP}(f)$ is the amplitude characteristic and $\varphi_{BP}(f)$ is the phase characteristic of the bandpass filter. $K_{F,BP}$ is equal to zero as the impact of the bandpass filter on both I and

Q path is the same. $A_{F,BP}$ is the input signal independent offset, which is assumed to be zero.

$$\begin{aligned} V_{F,BP} &= S_{BP}(f) = A_{BP}(f)\exp\left(j\varphi_{BP}(f)\right) \\ K_{F,BP} &= 0 \\ A_{F,BP} &= 0 \end{aligned} \tag{3.8}$$

The VKA model of the lowpass filter is presented in Eq. (3.9). $S_{LP,I}(f)$ and $S_{LP,Q}(f)$ are transmission S-parameters of the lowpass filter and baseband amplifier blocks for the I and Q path. $A_{LP,I}(f)$ and $A_{LP,Q}(f)$ are the amplitude characteristics of these amplifiers and the lowpass filter block I and Q paths accordingly. $\varphi_{LP,I}(f)$ and $\varphi_{LP,Q}(f)$ are the phase characteristics of the I and Q paths accordingly, see Eq. (3.10). $V_{F,LP}$ is the common part for both paths of filter characteristics, $K_{F,LP}$ represents the difference between both paths. $A_{F,LP}$ is assumed to be zero.

$$\begin{aligned} V_{F,LP} &= \frac{S_{LP,I}(f) + S_{LP,Q}(f)}{2} \\ K_{F,LP} &= \frac{S_{LP,I}(f) - S_{LP,Q}(f)}{2} \\ A_{F,LP} &= 0 \end{aligned} \tag{3.9}$$

$$\begin{aligned} S_{LP,I}(f) &= A_{LP,I}(f)\exp\left(j\varphi_{LP,I}(f)\right) \\ S_{LP,Q}(f) &= A_{LP,Q}(f)\exp\left(j\varphi_{LP,Q}(f)\right) \end{aligned} \tag{3.10}$$

3.3.3 IQ-Mixer

An idealized mixer can be modeled with the VKA-representation as described in Eq. (3.11). The f_c is the LO frequency.

$$\begin{aligned} V &= \exp\left(j2\pi f_c t\right) \\ K &= 0 \\ A &= 0 \end{aligned} \tag{3.11}$$

3.4 General Model of Nonidealities

A general model of nonidealities in a radar receiver is described in this section. The imbalance of amplitude and phase between the I and Q path, the coupling between the I and Q path as well as between different channels and DC offset is described in following. Each of the effects is described in the proposed VKA-representation in frequency domain.

3.4.1 Amplitude and Phase Imbalance

The imbalance between I and Q can be divided into an amplitude imbalance and a phase imbalance. The amplitude imbalance can be described as in Eq. (3.12). A corresponding VKA model of the amplitude imbalance is given in Eq. (3.13). a_{imb} is the amplitude imbalance between the I and Q path. This parameter can be frequency dependent or might be approximated as a constant value.

$$Y(f) = (1 + a_{imb}) \cdot \Re\{X(f)\} + (1 - a_{imb}) \cdot \Im\{X(f)\} \tag{3.12}$$

$$\begin{aligned} V_{F,amp} &= 1 \\ K_{F,amp} &= a_{imb} \\ A_{F,amp} &= 0 \end{aligned} \tag{3.13}$$

The phase imbalance is described with Eq. (3.14). The imbalance can also be easily described with the VKA-representation as is Eq. (3.15). φ_{imb} is the phase imbalance.

$$Y(f) = \exp(j\frac{\varphi_{imb}}{2}) \cdot \Re\{X(f)\} + \exp(-j\frac{\varphi_{imb}}{2}) \cdot \Im\{X(f)\} \tag{3.14}$$

$$\begin{aligned} V_{F,amp} &= \cos\left(\frac{\varphi_{imb}}{2}\right) \\ K_{F,amp} &= j\sin\left(\frac{\varphi_{imb}}{2}\right) \\ A_{F,amp} &= 0 \end{aligned} \tag{3.15}$$

In the general case the phase imbalance is frequency dependent. The imbalance effect can be simplified and assumed to be constant over the frequency in a narrow band system.

3.4.2 IQ Coupling

The VKA-representation gives a very good opportunity to implement in an easy and uncomplicated way the coupling between I and Q path. It may also be used to model the coupling between different receiver channels. The IQ coupling matrix C_{IQ} is described with Eq. (3.16). A symmetrical coupling between I and Q paths is assumed. The t_{IQ} and k_{IQ} are the transmission and the coupling parameter accordingly. The coupled and the transmitted power in the ideal case, with assumption of no losses, gives the input power, $t_{IQ}^2 + k_{IQ}^2$ equal to one.

$$C_{IQ} = \begin{bmatrix} S_{II} & S_{IQ} \\ S_{QI} & S_{QQ} \end{bmatrix} = \begin{bmatrix} t_{IQ} & k_{IQ} \\ k_{IQ} & t_{IQ} \end{bmatrix} \tag{3.16}$$

The VKA model of the coupling between I and Q paths is described with Eq. (3.17). $V_{coupling}$ is equal to the transmission parameter from the coupling matrix t_{IQ} and $K_{coupling}$ to the coupling parameter between paths I and Q.

$$\begin{aligned} V_{coupling} &= t_{IQ} \\ K_{coupling} &= k_{IQ} \\ A_{coupling} &= 0 \end{aligned} \tag{3.17}$$

In general case the t_{IQ} and k_{IQ} are frequency dependent.

3.4.3 Coupling between Receiver Channels

The coupling between the separate receive paths can be modeled in a similar way as the coupling between I and Q paths. The coupling matrix $S_{channel}$ describes the coupling of the signal from one receiver channel to another receiver channel. The dimension of the matrix is then N by N, where N is the number of receive channels.

$$S_{channel} = \begin{bmatrix} S_{11} & S_{12} & \dots & S_{1N} \\ S_{21} & S_{22} & \dots & S_{2N} \\ \dots & \dots & \dots & \dots \\ S_{N1} & S_{N2} & \dots & S_{NN} \end{bmatrix} \tag{3.18}$$

3.4.4 DC-Offset

The DC-Offset occurs in a mixer when part of the LO signal is coupled and mixed with itself. The DC-Offset can be easily modeled with the VKA-representation as in Eq. (3.19).

$$\begin{aligned} V_{DC} &= 1 \\ K_{DC} &= 0 \\ A_{DC} &= a_{DC} \exp(j\varphi_{DC}) \end{aligned} \tag{3.19}$$

a_{DC} is the amplitude and the φ_{DC} is the phase of the DC-Offset. The DC-offset in VKA-representation is a complex value. The I and Q paths are described with a single parameter. The phase described the difference between the I and Q paths.

3.5 Measurements of the HRWS SAR Demonstrator

The direct down-conversion receiver architecture has been chosen for the HRWS SAR demonstrator designed and made by Astrium GmbH. The demonstrator board, which can be seen in Fig. 3.1, consists of eight independent receive channels. Signals from each channel are quantized with one bit. The board demonstrates the low bit multi-channel radar principle. More about multi-beam low-bit quantization can be found in Chapter 5. The HRWS SAR demonstrator was used as a model for the receiver in the simulation. Components of the receiver front-end of the demonstrator were measured in order to provide more accurate data for the simulation. Characteristics presented in following paragraphs were measured on one of the eight receive channels of the demonstrator. Fig. 3.6 shows a diagram of a single channel of the demonstrator board. The amplifiers and the lowpass filter for I and Q band were built as a single block. In this case the measurements considered a block consisting of two amplifiers and a lowpass filter. The characteristic of the IQ-mixer in the receiver front-end have been taken from data sheets of this component.

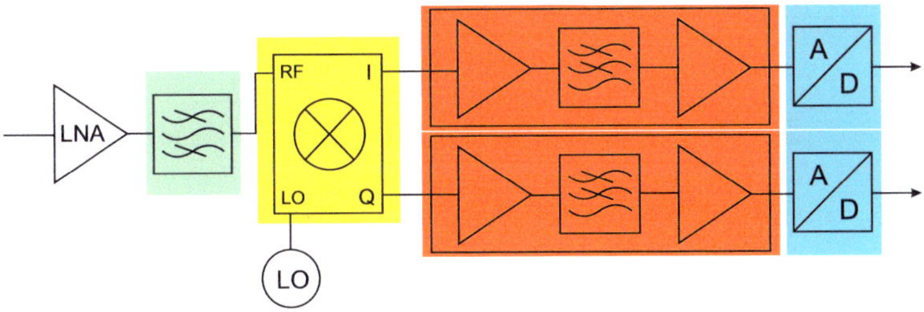

Figure 3.6: Diagram of the direct down conversion receiver, single receive channel of the HRWS SAR demonstrator.

3.5.1 Bandpass Filter

The amplitude and the phase characteristic of the measured bandpass filter can be seen in Fig. 3.7. The pass band at about 10 GHz can be clearly distinguished in Fig. 3.7(a) showing the amplitude characteristic. Fig. 3.7(b) shows the phase characteristic of the measured bandpass filter. The phase characteristic is linear in the pass band.

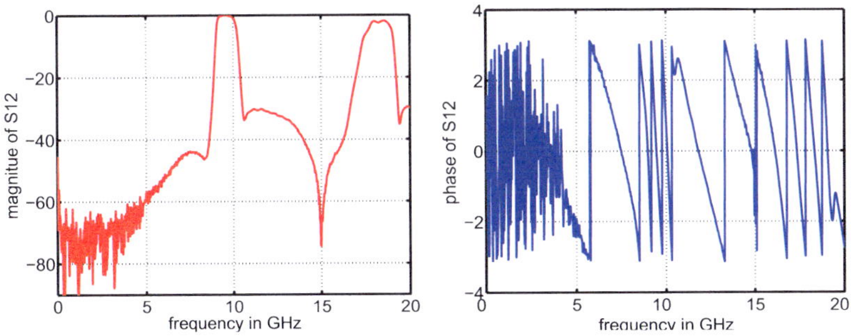

(a) Amplitude characteristic

(b) Phase characteristic

Figure 3.7: Measured characteristics of the bandpass filter.

The $V_{F,BP}$ parameter is equal to the measured characteristic. The $K_{F,BP}$ and $A_{F,BP}$ are equal to zero.

3.5.2 Amplifier and Lowpass Filter Chain

The I and Q receive paths were measured separately with a network analyzer. In the demonstrator amplifiers and lowpass filters were designed and built as single block. Because of this the whole block was measured together for the I and the Q path as is indicated in Fig. 3.6 with orange rectangles. The characteristics for both modules for the quadrature and the inphase part of the signal can be seen in Fig. 3.8 and Fig. 3.9 accordingly. The maximal frequency of the whole module is about 400 MHz. The bandwidth of the HRWS SAR transmit signal is about 300 MHz. The phase characteristic of the module is linear for the pass area.

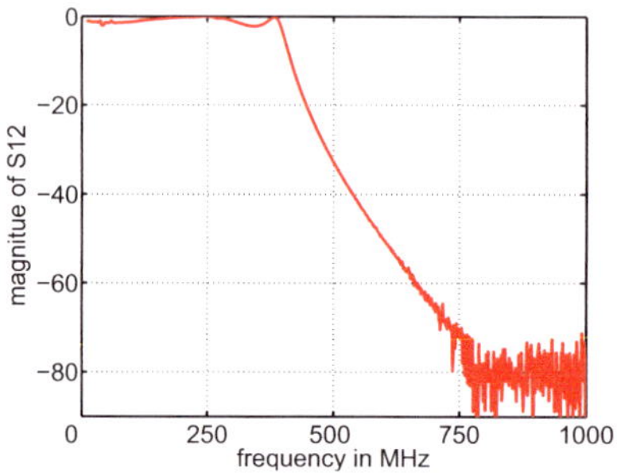

(a) Amplitude characteristics

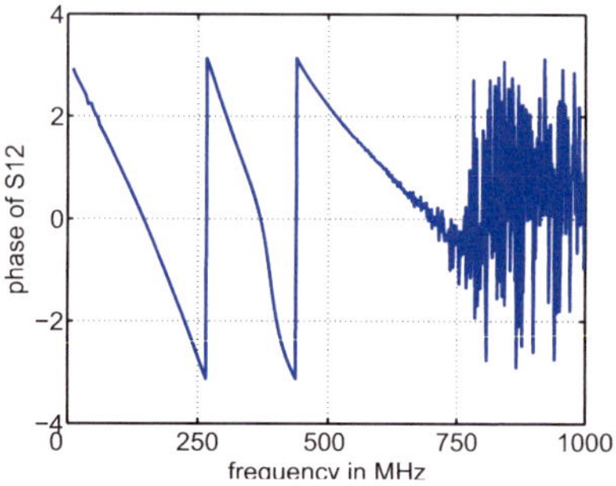

(b) Phase characteristic

Figure 3.8: Measured characteristics of the I receive path.

Fig. 3.10 and Fig. 3.11 show parameters $V_{F,LP}$ and $K_{F,LP}$ for lowpass filters and baseband amplifiers blocks of the HRWS SAR demonstrator. The $V_{F,LP}$ shows the common part for both I and Q paths. The magnitude of the parameter looks similar to characteristics of the components. The phase of the parameter is linear. The $K_{F,LP}$ shown in Fig. 3.11 represents the difference between the I and Q path.

3.5.3 Mixer

A MIQ66 mixer from Spectrum Microwave was used in the HRWS SAR demonstrator. Fig. 3.12 shows the imbalance in amplitude and quadrature phase according to data sheets of the IQ mixer MIQ6xSMD-1 used in the

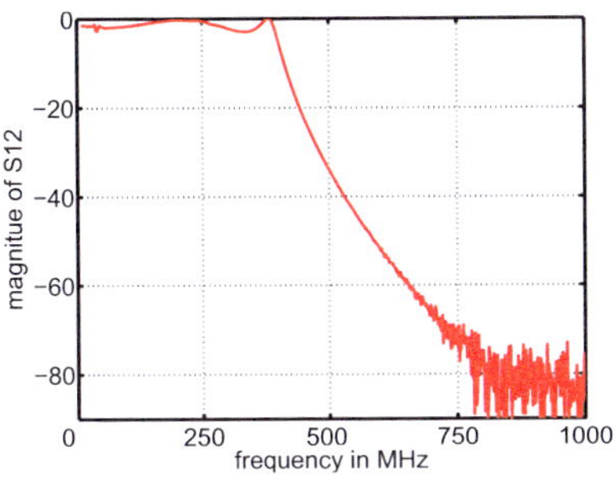

(a) Amplitude characteristic

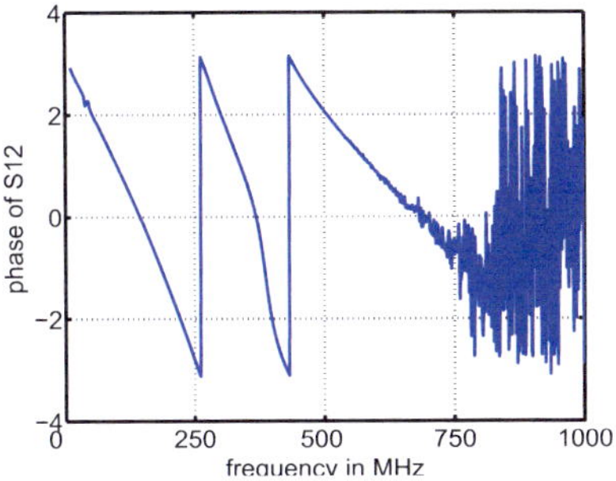

(b) Phase characteristic

Figure 3.9: Measured characteristics of the Q receiver path.

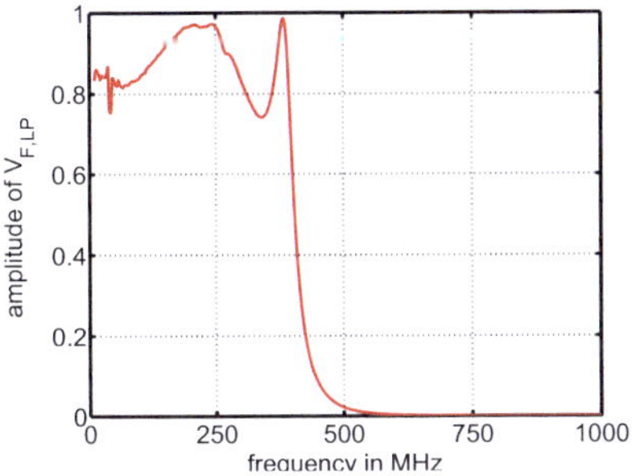

(a) Amplitude of $V_{F,LP}$ as a function of frequency

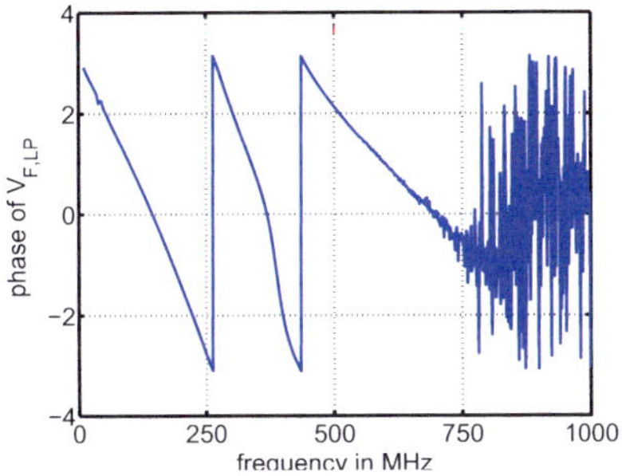

(b) Phase of $V_{F,LP}$ as a function of frequency

Figure 3.10: The $V_{F,LP}$ for the baseband amplifiers and lowpass filter calculated from measurements of the HRWS SAR demonstrator.

HRWS SAR demonstrator. The imbalance can also be modeled with the VKA-representation.

3.5.4 Analog-to-Digital Converter (ADC)

The HRWS SAR demonstrator was built with one bit quantization ADCs at each of the eight channels. The output of a channel depends on the input

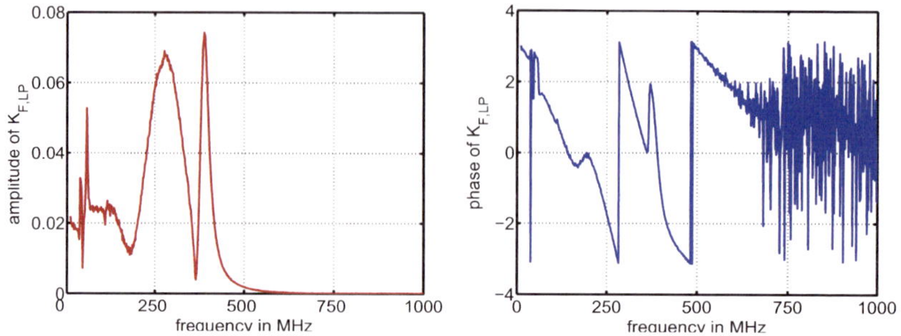

(a) Amplitude of $K_{F,LP}$ as a function of frequency.

(b) Phase of $K_{F,LP}$ as a function of frequency

Figure 3.11: The $K_{F,LP}$ for the baseband amplifiers and lowpass filter calculated from measurements of the HRWS SAR demonstrator.

signal and can be high or low, which is defined as zero and one or -1 and 1. Signals of six out of eight one bit receive channels were measured. The eighth channel was used as a reference with an eight bit ADC. During measurements a DC offset between I and Q path of each of the ADC converters has been observed. Fig. 3.13 shows the result of measurements of six channels of the demonstrator's ADC for both I and Q paths. In this figure plotted are average values for each of the channels for both I and Q paths. The 'low' state is here defined as zero and the 'high' state as one. The mean value of the signal with no DC offset is 0.5 in the ideal case. In Fig. 3.13 three channels show mean values above 0.5 and three below 0.5 value. This gives a different DC offset on each of observed channels. In the figure it should be also noted that these mean values differ slightly for I and Q path of the same channel. In the demonstrator the threshold for the channel can be adjusted in a particular area for I and Q. The adjustment is the same for both the I and Q paths as it was realized with a single component. The difference in the threshold for the I and Q paths could not be adjusted separately. The difference between the I and Q path threshold is constant for a specific ADC block.

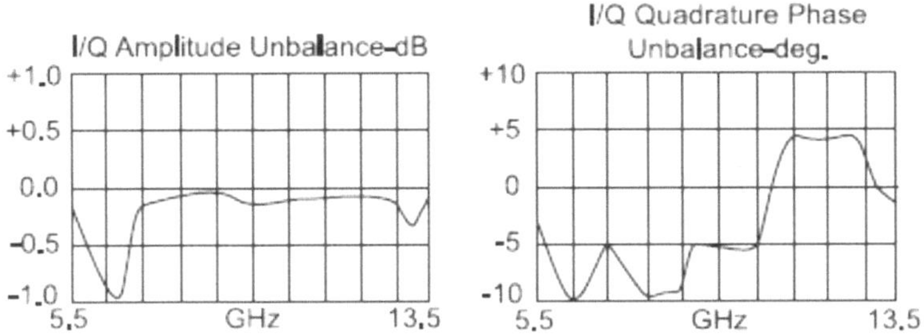

Figure 3.12: Amplitude and quadrature phase imbalance from MIQ6xSMD-1 IQ mixer data sheet.

3.6 Simulation

A simulation of the HRWS SAR was developed in MatLab. The simulation is based on a system with a single transmit antenna and multiple receive apertures. The position of all apertures can be chosen freely in three dimensional space. Simulated is only a single transmit signal. This is sufficient to determine the influence of receiver nonidealities, quantization and antenna position on the performance of the system in elevation (range). The flowchart of the simulation can be seen in Fig. 3.14. The simulation includes the model of the receiver front-end and possible nonidealites that can occur [63]. The simulation is divided into two parts to reduce the computation time by series of simulation with various parameters. In the first part the receive signal for each sub-aperture is generated. The number and the position of all antennas, the scenario and the acquisition geometry must be defined for the first part of the simulation. The result is a time signal generated for each of the receive sub-apertures. All signals are saved in separate files. The second part of the simulation includes additive noise, transmitter front-end model and quantization. Signals from all receive sub-apertures are then processed with the DBF scan-on-receive algorithm described in Section 2.5.1. The radar data is compressed with a SAR processing algorithm.

This simulation is also used in Chapter 4 and Chapter 5, [61], [64]. The simulation generates the radar data for investigation of the influences of minor

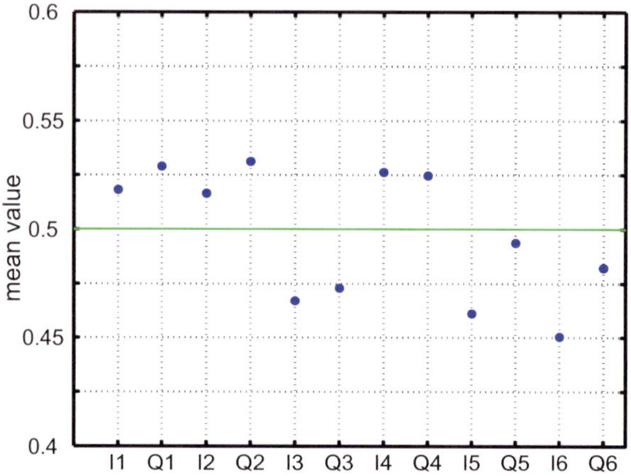

Figure 3.13: DC offset estimated for six of the channels of the HRWS SAR demonstrator for I and Q paths.

position errors of receive antenna array element on SAR performance in elevation in Chapter 4. In Chapter 5 the simulation provided results for radar data with various number of quantization bits.

3.7 Simulation Results

Results of the simulation with chosen nonidealities are presented in this section. The simulated system consists of $N = 60$ independent receive subapertures in elevation and a single one in azimuth. This corresponds to a single panel of the HRWS SAR receive antenna array. The simulated system transmits a signal with a bandwidth B of 300 MHz. Time duration τ of the transmitted signal is set to 70 μs. The radar platform is flying on a round orbit at an altitude of about 700 km. The incidence angle is set to 30°. A single point target scenario is simulated for each example. Simulated was the influence of the amplitude imbalance and phase imbalance.

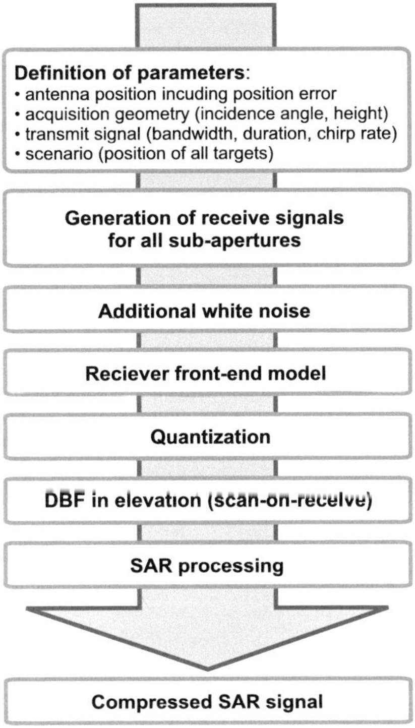

Figure 3.14: Flowchart of the direct simulation.

3.7.1 Amplitude Imbalance

The influence of the amplitude imbalance was simulated for various values of the parameter a_{imb}. Simulations were conducted for an input signal-to-noise ratio of 10 dB, 0 dB and -10 dB. The amplitude imbalance a_{imb} was set to 0, 0.05, 0.1, 0.2, 0.3, 0.4 and 0.5. Fig. 3.15(a) presents simulation results for the SNR of 10 dB. Fig. 3.15(b) and Fig. 3.15(c) show simulation results for the SNR of 0 dB and -10 dB accordingly. As it can be seen in all of those figures with the increasing amplitude imbalance the noise level rises. Fig. 3.15(d) shows the change of the noise level for all simulated values of the amplitude imbalance for the three values of the input SNR.

In Fig. 3.12 presented is the amplitude and phase imbalance of the MIQ6xSMD-

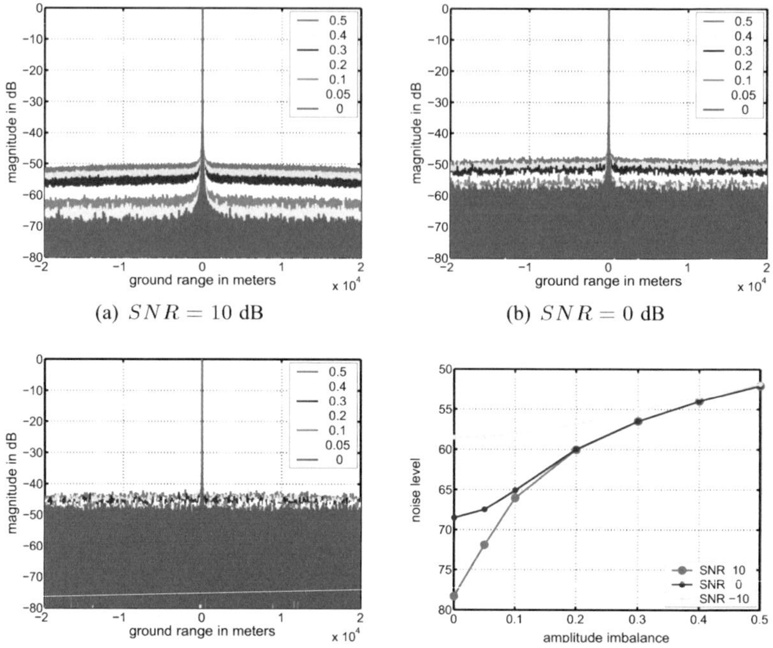

(a) $SNR = 10$ dB (b) $SNR = 0$ dB

(c) $SNR = -10$ dB (d) Noise level

Figure 3.15: Simulation results for the amplitude imbalance (amplitude imbalance linear)

1 IQ mixer. The maximal amplitude imbalance between I and Q path for this mixer according to the data sheets is about 1 dB, which corresponds to a_{imb} of about 0.05. This amplitude imbalance increases the relative noise floor level according to simulations by about 5 dB assuming input SNR of 10 dB.

3.7.2 Phase Imbalance

The phase imbalance was simulated for various values of the parameter φ_{imb}. The input signal-to-noise ratio was set to 10 dB, 0 dB and -10 dB. Fig. 3.16 shows results for simulations for φ_{imb} of $0°$, $5°$, $10°$, ... , $35°$. Fig. 3.16(a) shows simulation results for SNR of 10 dB, Fig. 3.16(b) for SNR of 0 dB and Fig. 3.16(c) for SNR of -10 dB. With increasing phase imbalance the relative noise floor level increases. This is clearly visible in figures showing simulated signals. Fig. 3.16(d) shows the change in the noise floor level for all simulations. In this figure plotted are also results of simulations for a phase imbalance between $1°$ and $10°$ with a step of $1°$.

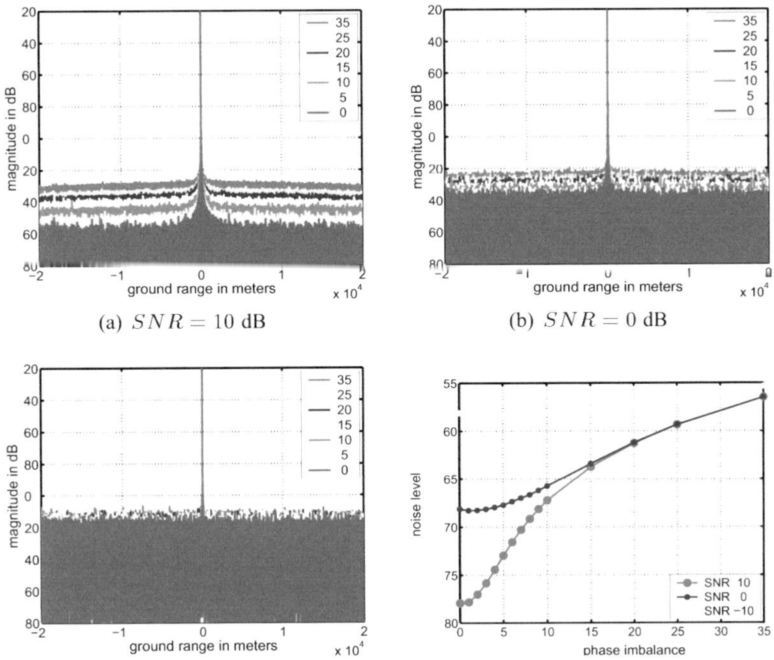

(a) $SNR = 10$ dB (b) $SNR = 0$ dB

(c) $SNR = -10$ dB (d) Noise level

Figure 3.16: Simulation results for the phase imbalance (phase imbalance in degrees)

Fig. 3.12 shows the imbalance characteristic for the MIQ6xSMD-1 IQ mixer. Te maximal phase imbalance for this mixer type is between $-10°$ and $5°$. According to simulations for a phase imbalance of $5°$ the noise floor increases by about 5 dB, for $10°$ the noise floor increases by about 11dB assuming input SNR of 10 dB.

3.8 Summary

The simulation of the radar receiver front-end and various nonideality effects were described in this chapter. The direct down-conversion receiver was chosen as a model. The same type of a receiver was used in the HRWS SAR demonstrator built by Astrium GmbH. To describe the components of the direct down-conversion receive a VKA-representation in frequency domain war proposed. A new representation has the capability to describe all receiver components as well as some effects like e.g. coupling between the I and Q receiver paths. The proposed representation is an extension of VKA-representation described in [48] in the frequency domain. It was showed that the representation provided an easy and flexible way to describe components of the receiver front-end. The possibility to use one model to describe at the same time the I and Q path after demodulation is of advantage for simulation. Each of the receiver components or blocks of components can be described with the three parameters of the VKA-representation. Also effects in receiver chain like e.g. amplitude and phase imbalance, coupling between I and Q path and DC offset can be described in the same way. The advantage of the representation is the possibility to describe all receiver components and nonidealities effect in a consistent way. All the components of a direct down-conversion receiver were recalculated to the proposed VKA-representation. Also the amplitude and phase imbalance, coupling and DC offset were represented in the same way. Components of the HRWS SAR demonstrator had been measured and converted into the VKA-representation. The amplitude and phase imbalance was simulated to show the capability of the implemented VKA-representation simulation. Various values of the amplitude and phase imbalance were simulated with three values for signal-to-noise ratio of the radar input signal.

4 Antennas

A basic radar system has a single antenna, which is used for transmitting and receiving radar signals. This type of antenna system is called a monostatic configuration. More complex systems can have two separate antennas – a bistatic configuration, where one antenna is used for transmit and the second one for receive, or multiple antennas for transmit and receive – a multistatic configuration. Antennas can be mounted on a single platform or on multiple platforms flying in a constellation. In this chapter considered are possible minor mechanical distortions that can occur in an antenna system. An antenna configuration used in the High-Resolution Wide-Swath (HRWS) SAR was taken as a model for simulations. The antenna system of the HRWS SAR consists of a single transmit antenna and an N by M receiver antenna array with independent receive channels. The receive front-end was described in previous chapter. The DBF is applied in both range and azimuth direction. In this work considered is the DBF in range direction. Minor position errors may occur in the antenna array being caused by e.g. mechanical stress, production or assembly errors or failures. The influence of those minor position errors on the performance of the SAR system is investigated in this chapter.

Various possible geometric configurations of antenna arrays in a SAR system are briefly described in Section 4.1. The HRWS SAR antenna configuration is described in this section in more detail. This configuration is used as a basis for the modeling of minor position errors in Section 4.2. A mathematical model for the array factor and the antenna characteristic is presented for the ideal linear array configuration and for the same linear array with position errors in Section 4.3. Simulations of the influence of the minor position errors are described in Section 4.4. First the antenna array factor for a linear array is simulated. In following subsection this antenna array is used for simulation of a SAR image. The influence of a disturbed antenna array on the resulting radar image is investigated.

4.1 Geometry of a SAR Antenna System

A monostatic SAR system consisting of a single radar antenna is presented in Fig. 4.1. In this configuration a single antenna is used for transmit and for receive. With increasing demand on radar performance new solutions were developed, which introduced also bistatic and multistatic antenna configurations. The separation of the transmit and receive antenna gives more degrees of freedom, which can be used to improve the flexibility of the system and provide more operational modes of the radar system and improve its performance. In bistatic and multistatic systems transmit and receive antennas can be mounted on the same, on two or in case of a multistatic system on many separate platforms. The separation of the transmit and receive antenna allows for optimization of each separately. A multistatic SAR system consists of multiple transmit and/or receive antennas. An example of such antenna configuration is presented in Fig. 4.2 with two transmit antennas marked with red color and four receive antennas marked with blue color. Some antennas in such configuration can also be used for both transmit and receive. A good example of such system with a transmit/receive antenna and a second receive only antenna is the Shuttle Radar Topography Mission (SRTM) providing radar interferometry[1] with a second receive antenna [74]. In the shuttle mission SRTM/X-SAR a single pass SAR interferometry has been achieved through a simple antenna system: a primary transmit/receive antenna system is mounted on board of the shuttle and a second receive-only antenna is mounted outboard on a 60 meter boom oriented in the cross-track direction. The off-board antenna with the inboard system is forming a parallel antenna system, [42]. More information about the SRTM/X-SAR system can be found in [16].

An application example of a multistatic antenna configuration is the HRWS SAR. The HRWS SAR antenna system is a multistatic configuration with a single transmit antenna and a receive array. The configuration is presented in Fig. 4.3. Each antenna array element has its independent receive front-end so that a separate receive signal is provided from each receive sub-element. The receive antenna array elements are aligned, as can be seen in Fig. 4.3, in M panels (azimuth direction) each with N independent receive elements

[1] Interferometry refers to methods, where superimposition of waves is used to extract information. The interferometry is a common technique used e.g. in astronomy, remote sensing and optics.

Figure 4.1: Monostatic configuration.

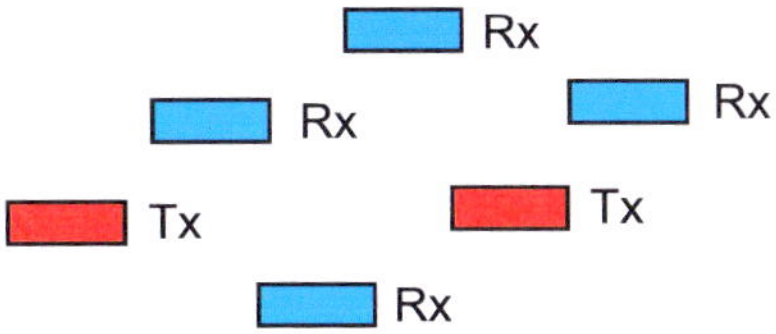

Figure 4.2: Multistatic configuration with multiple Tx and Rx antennas.

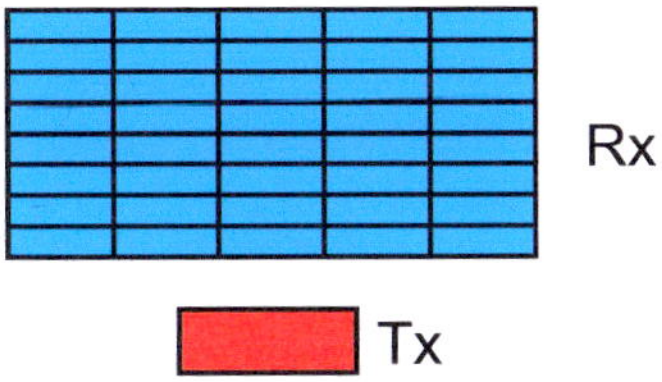

Figure 4.3: Antenna configuration in the HRWS SAR.

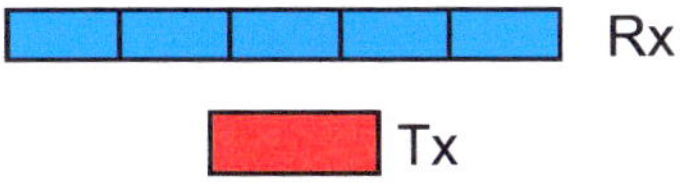

Figure 4.4: Multistatic configuration for DBF in azimuth.

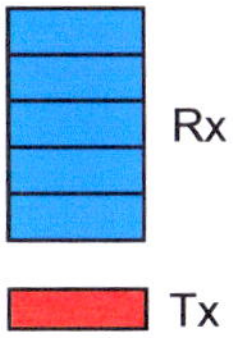

Figure 4.5: Multistatic configuration for DBF in elevation/range.

(elevation/range direction). The DBF and the SAR processing are linear operations, therefore properties of the antenna array in range and azimuth can be analyzed separately. Fig. 4.4 shows an antenna array configuration with the DBF in azimuth direction only. Fig. 4.5 shows the antenna array configuration with the DBF in range direction only. The configuration for the DBF in range only is considered in more detail in following sections.

4.2 Geometry of Minor Position Errors

This work concentrates on the influence of minor position errors of antenna array elements on the performance of the SAR system with the DBF in range within a single panel. A schematic model for position errors of receive array elements is shown in Fig. 4.6, [60], [62]. Position errors considered in the simulation are in slant range direction. The maximal minor position error for an array is defined by dx_{max}. The distance between two neighboring array elements is h_s. The SAR system platform is flying at the altitude R_{Ground} above the ground. Minor position errors are indicated in the figure with the red dashed line with a maximal misplacement of dx_{max}. The position of each of the sub–apertures is calculated from the middle of the array. In the figure d_1 is the distance to the middle of the array for the first sub-aperture, d_2 for the second and so on.

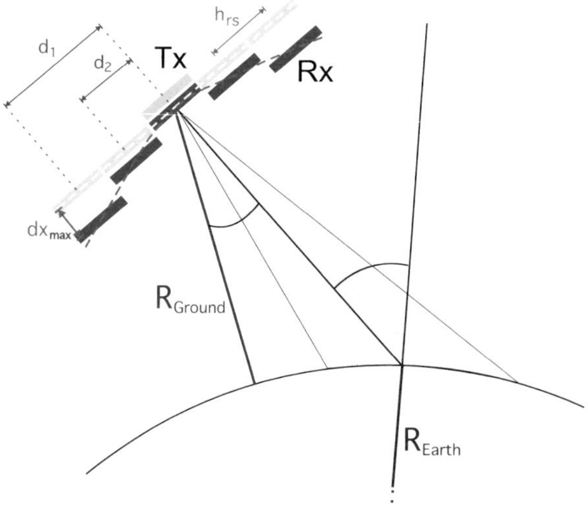

Figure 4.6: Schematic model of the minor antenna array element position errors of the radar system.

A single panel is sufficient to estimate the performance as only the range direction is considered. This reduces significantly the number of elements of the antenna array to be simulated from M by N to 1 by N. It is assumed that all panels have the same radiation characteristic. In this case results obtained from the simulation of a single panel can be generalized for all receive panels. In the simulation the transmit and the receive antennas are positioned on the same platform, so that there are no distance, velocity or alignment errors caused by a different movement of separate platforms. In the simulation a virtual transmit antenna is positioned in the middle of the receive antenna array, as shown in Fig. 4.6. The receive signal is calculated for each of the sub-apertures. The distance between the transmit aperture and a simulated point target positioned on the illuminated area is calculated. Then also the distance between each simulated point target and each sub-aperture is calculated. The delay of the receive signal for each sub-aperture can be calculated from the overall distance the transmitted signal travels. The radar receive signal in case of multiple scatterers is a superposition of accordingly delayed transmit signals reflected on each target. These delays are calculated from the known distance of the receive sub-aperture phase center to the middle of the array d_n, distance to the ground R_{Ground} and the middle Earth radius R_{Earth}, incidence angle φ_i, NaDir angle β and position of point targets on the ground.

4.3 Influence of Position Errors on Radiation Pattern of the Antenna Array

The HRWS SAR receiver antenna array has been designed as a regularly-distributed planar array. All array elements have the same radiation pattern. The combined radiation pattern of all sub-apertures creates the array beam pattern. The resulting beam pattern of the antenna array can be calculated as a multiplication of the array factor (AF) with the beam pattern of a single array element, see Eq. (4.1), [88].

$$I(\theta) = AF(\theta) \cdot I_0(\theta) \tag{4.1}$$

Where $I(\theta)$ is the resulting antenna array pattern, $I_0(\theta)$ is the pattern of a single element and θ is the angle. The array factor is described with $AF(\theta)$. The array factor of an N-element linear array with a spacing between element and the middle of the array d_n for the nth element is represented by Eq. (4.2).

$$AF(\theta) = \sum_{n=1}^{N} \exp(j\Psi_n(\theta)) \tag{4.2}$$

$$\text{Where} \quad \Psi_n(\theta) = kd_n \cos(\theta) \tag{4.3}$$

$$k = \frac{2\pi}{\lambda} \tag{4.4}$$

Ψ_n is the difference in phase between the nth sub-aperture in the receive array and an element positioned in the middle of the array[2] and k is the wave number. Eq. (4.2) can be simplified, assuming a regular distribution of antenna array elements in a $\lambda/2$ distance, to Eq. (4.5).

$$AF(\theta) = \frac{\sin(N \cdot \frac{\theta}{2})}{\sin(\frac{\theta}{2})} \tag{4.5}$$

A model for the minor position error of a single element in the antenna array is presented in Fig. 4.7. Three consecutive antenna array elements, each marked with a point, are presented in this figure. In the figure marked are only phase centers of array elements. The reference point for the phase difference of the array element is the utmost right element shown in the figure. Other signals are regarded relative to the signal received by this utmost right sub-element. The middle element is misplaced in a plane perpendicular to the antenna array plane. The distance of the middle array element relative to the right element is described with d. The signal is received from an angle θ. Eq. (4.3) describes the phase difference between the middle and the right array element for the ideal not disturbed antenna array.
The middle element is misplaced by a position error dx, as shown in Fig. 4.7. The distance between elements changes with the misplacement of the array element. Also the phase difference Ψ to the upmost right element for the misplaced element changes. To estimate the error caused by element misplacement the distance d' is calculated. First the misplaced array element is projected on the array plane as shown in Fig. 4.7. A line perpendicular to the

[2] In case of an even number of array elements the phase difference is calculated between the nth array element and the one in the middle. For odd numbers of elements the difference is calculated between the nth element and a synthetic element positioned in the middle of the array.

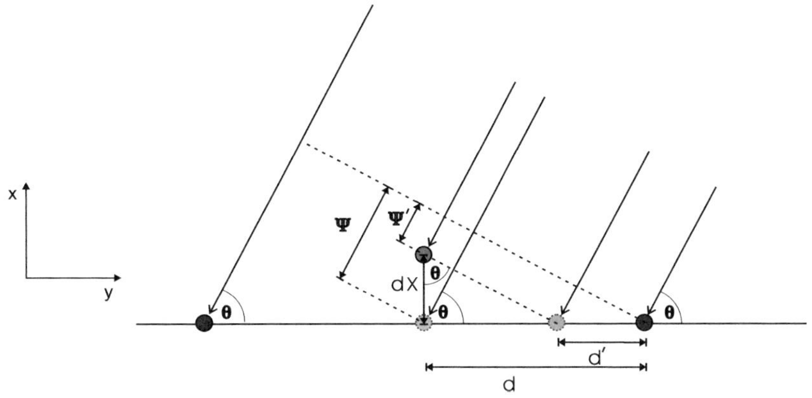

Figure 4.7: Schematic antenna array with misplaced middle element.

direction of the incoming wave is drawn through the phase center of the misplaced element. The cut point of this line with the array plane is marked in the figure with an orange point. The new distance between the first element and the projection of the middle element on the array plane d' can be calculated according to Eq. (4.6).

$$d'(\theta) = d - dx \cdot \text{ctan}(\theta) \tag{4.6}$$

It should be noted that d' depends on the angle of the incoming wave θ. The phase difference Ψ' for the middle element with position error can be calculated with known d' according to Eq. (4.7).

$$\Psi'(\theta) = \frac{2\pi}{\lambda} kd'(\theta)\cos(\theta) = \frac{2\pi}{\lambda}\left[\cos(\theta)k\left(d - dx \cdot \text{ctan}(\theta)\right)\right] \tag{4.7}$$

In this example the phase difference for the misplaced array element is smaller than it would be in the ideal case without the position error. For distortion in the opposite direction, giving a misplacement of $-dx$, the phase difference would be higher than in the non distorted case. The phase error caused by the misplacement can be calculated according to Eq. (4.8).

$$\Psi_{error}(\theta) = \Psi'(\theta) - \Psi(\theta) = -\frac{2\pi}{\lambda}\left[\cos(\theta)dx \cdot \text{ctan}(\theta)\right] \tag{4.8}$$

The array factor of the distorted array can be calculated according to Eq. (4.9)

$$AF(\theta) = \sum_{n=1}^{N} \exp(j\Psi'(\theta)) \tag{4.9}$$

In Eq. (4.9) the combined beam of the antenna array is not focused correctly because of phase errors. In this case the array is not using its optimal gain for receive. The radiation pattern of the array can be disturbed. The combined beam may also be slightly shifted in the looking direction so that the main lobe is not exactly at the expected angle.

4.4 Simulation Results of a Receive Antenna Array with Position Errors

The influence of the minor position error on antenna array, antenna pattern and the SAR system performance is analyzed in this section. Results of simulations are presented and compared. The HRWS SAR antenna configuration is used as an antenna array model. In the HRWS SAR system the receive antenna is designed as an array with many sub–apertures. Although the size of a single sub–element is rather small, the size of the whole array can be large. In simulations the number of antenna array elements in elevation N has been set to 60. The height of one sub-element h_{rs} is 16.7 mm, so that the whole antenna array with 60 elements in elevation has about 1 meter length. The distance between elements d is 16.7 mm. The radar is working on a middle frequency of 9.65 GHz. The wavelength λ in the middle of the band is 31.1 mm. The transmitted signal has a bandwidth of 300 MHz. Fig. 4.8 shows two types of deformations of the antenna array panel that are considered in this work. One possible deformation of the antenna plane is a bend in the shape of a parable, [9], [93], which is presented in Fig. 4.8(a). This type of distortion can be approximated as a symmetrical bend of array's edges. Further in this chapter this type of antenna array distortion will be referred to as a symmetrical bend. The second possibility is a bend, where array edges are bent in opposite directions as shown in Fig. 4.8(b). As this distortion has no symmetry on XY-plane, it is referred further as an unsymmetrical bend. In Fig. 4.8 the maximal position error dx_{max} is marked for both types of bend. Simulations have been performed for dx_{max} between 0 mm (no error) and 15 mm. The bend was modeled as a quadratic function of the distance calculated from the middle of the array.

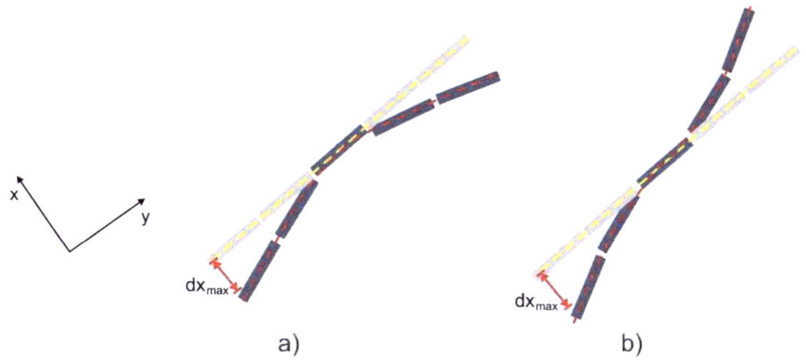

Figure 4.8: Position errors of the receiver antenna array elements (a) symmetrical distortion (b) unsymmetrical distortion.

The influence on the array directivity for both models of the position error is investigated in Section 4.4.1. The simulation is based on Eq. (4.9), which includes a position error for each of the antenna array elements. The HRWS SAR antenna array is used as a model for the simulation. The influence of the minor position errors on the performance of a SAR system is presented in Section 4.4.2. The SAR simulation as described in previous chapter is used to obtain results. The minor position errors applied to the antenna array are the same as in Section 4.4.1.

4.4.1 Array Directivity

Results of simulation for the array directivity are presented in this section. The results are obtained according to Eq. (4.9) with various values of the maximal position error dx_{max}. An regularly distributed antenna array with 60 elements is simulated. Those elements are positioned in a separation distance of $d = 16.7$ mm with the height of a single element of $h_{rs} = 16.7$ mm. The antenna array beam is focused at $\theta = -1°$. The antenna pattern of a 60 element array with a symmetrical position error is presented in Fig. 4.9. The bend of the antenna was simulated for a maximal error dx_{max} of 0 to 15 mm with a step of 1 mm. The black curve marks the reference simulation without distortion of the array plane (dx_{max} equals to 0 mm). Other curves show results of simulations for a maximal position error dx_{max} of $1, 2, \ldots, 15$ mm. All antenna patterns are scaled so that the maximal value of the ideal curve is at 0 dB.

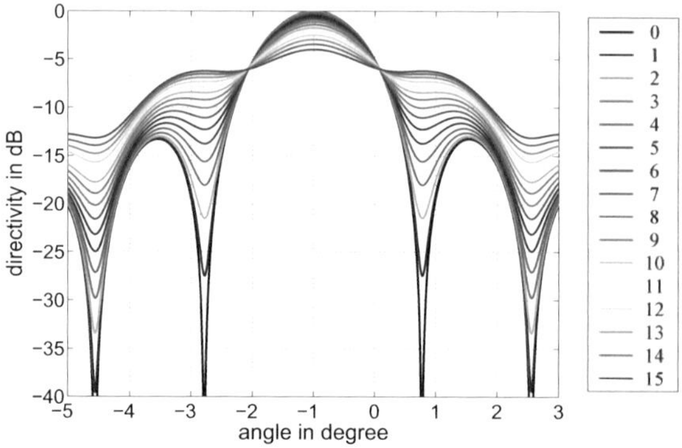

Figure 4.9: Array Factor with symmetrical position error for dx_{max} of $0, 1, 2, \ldots, 15$ mm.

The influence of the symmetrical position error of antenna elements is presented in Fig. 4.9. Main effects are the decrease of the antenna array gain at $-1°$ (main lobe), broadening of the main lobe (3 dB bandwidth) and the decrease of the peak to side lobe ratio (PSLR). Values of the loss of gain, the 3 dB bandwidth of the main lobe and the PSLR for each simulation presented in Fig. 4.9 are collected in Table 4.1. It is to be noted that the symmetrical bend of the array does not influence the position of the main lobe. The maximum of the main lobe is for all simulated arrays exactly at $-1°$.
The simulation was repeated for the unsymmetrical antenna array bend configuration. Fig. 4.10 shows results for the antenna array pattern with the unsymmetrical position error. The black curve in the figure marks the ideal antenna array with no position errors. All simulation curves have been scaled so that the maximum of the ideal one is at 0 dB. The maximal position error dx_{max} was set to $1, 2, \ldots, 15$ mm, as in previous simulation for symmetrical bend. In Table 4.2 results for each simulated antenna array for the unsymmetrical distortion can be found. The table includes the decrease of the antenna array gain for the main lobe, the position of the main lobe in degrees, the broadening of the main lobe, the decrease of the PSLR and the loss of gain

dx_{max} in mm	Gain loss in dB	3 dB beam width in degree	PSLR in dB
0	0	1.57	13.25
1	0.02	1.57	13.16
2	0.07	1.58	12.89
3	0.15	1.59	12.46
4	0.27	1.60	11.88
5	0.42	1.61	11.18
6	0.61	1.63	10.38
7	0.84	1.65	9.51
8	1.10	1.69	8.59
9	1.39	1.73	7.64
10	1.73	1.78	6.73
11	2.10	1.86	5.73
12	2.52	1.97	4.80
13	3.00	2.16	3.91
14	3.47	2.71	3.05
15	4.01	4.66	2.21

Table 4.1: Results for symmetrical antenna array bend (AF)

at $-1°$ direction. In case of the unsymmetrical bend it is to be observed not only the loss in the main lobe but also a shift of the main lobe. The main lobe of the undisturbed array is at $-1°$, for maximal position error of 2 mm the maximum of the main lobe is at $-0.8°$, for a position error of 5 mm the maximum is at $-0.5°$, for maximum position error of 10 mm at $-0.1°$ and for 15 mm maximal position error at about $0.3°$. This effect can be clearly seen in Fig. 4.10. The loss of the antenna array gain for the main lobe is lower than in corresponding simulations for symmetrical antenna bend. In SAR system with the scan-on-receive DBF important is the gain loss in the direction the radar is focusing. The antenna gain loss at $-1°$ is higher for the unsymmetrical bend than in symmetrical bend case.

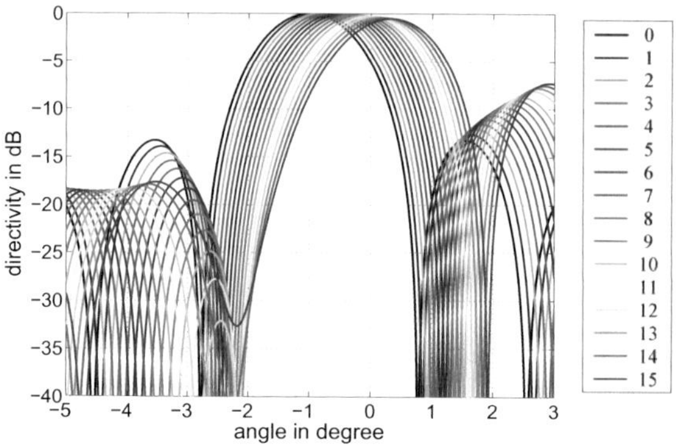

Figure 4.10: Antenna Factor with unsymmetrical position error for dx_{max} of $0, 1, 2, \ldots, 15$ mm.

4.4.2 Synthetic Aperture Radar with Minor Position Errors

The same antenna array configurations with symmetrical and unsymmetrical minor position errors of elements presented in Fig. 4.8 is used in a simulation of a radar system. Results for a satellite SAR system with an antenna array of 60 elements in elevation are presented in this section. The satellite platform, on which the radar is mounted, is flying at an 780 km altitude. The incidence angle has been set to 30°. The simulated signal is a chirp with a bandwidth B of 300 MHz and a time duration τ of 70 μs. The simulation used in Chapter 3 was adapted for modeling of minor position errors in the receive antenna array. The possibility to define a minor position error independently for each of antenna elements in direction corresponding to the x direction in Fig. 4.8 was implemented. The receive signal is calculated for each of antenna elements. Then the receive data is processed with the DBF scan-on-receive algorithm (see Section 2.5.1) and compressed with a SAR processing algorithm, in this case with the chirp scaling algorithm. The simulated scenario consists of a

dx_{max} in mm	Gain loss in dB	position of max. in deg.	3 dB beam width in degree	PSLR in dB	loss at -1 deg. in dB
0	0	-1	1.57	13.25	0
1	0	-0.91	1.57	12.65	0.04
2	0.01	-0.82	1.57	12.09	0.15
3	0.02	-0.73	1.57	11.56	0.34
4	0.04	-0.64	1.57	11.05	0.61
5	0.06	-0.56	1.58	10.58	0.95
6	0.09	-0.47	1.58	10.13	1.37
7	0.12	-0.38	1.58	9.70	1.87
8	0.15	-0.29	1.58	9.28	2.45
9	0.19	-0.20	1.58	8.89	3.10
10	0.24	-0.11	1.58	8.51	3.83
11	0.29	-0.02	1.58	8.14	4.64
12	0.34	0.07	1.59	7.78	5.51
13	0.40	0.14	1.59	7.44	6.46
14	0.47	0.23	1.59	7.10	7.46
15	0.54	0.32	1.59	6.78	8.50

Table 4.2: Results for unsymmetrical antenna array bend (AF)

single point target positioned on the ground. The target is at $-1°$ from the line of sight of the radar antenna array.
The distortion of the array has been set the same as in simulations of the array factor for symmetrical and unsymmetrical bend. Simulations have been carried out for the maximal position error dx_{max} of $0, 1, 2, \ldots, 15$ mm. The results for the symmetrical bend are presented in Fig. 4.11. To be noted is that the x-axis is in ground range in meters. The black curve corresponds to the ideal case with no antenna deformation. All curves are scaled so that the

maximum power of the ideal curve is at 0 dB. The decrease of the maximal power of the target can be clearly seen on this figure. The maximal power decreases accordingly to the maximum position error of the antenna array. Results for the unsymmetrical bend are shown in Fig. 4.12. The black curve is

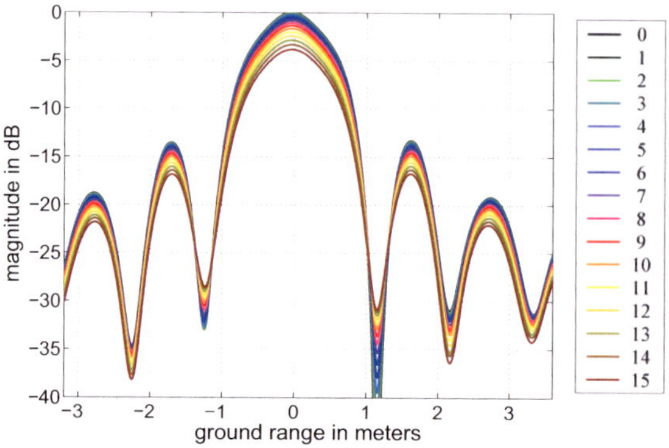

Figure 4.11: Simulation results for SAR with the DBF in range with symmetrical position errors.

corresponding with the ideal case. All curves are scaled so that the maximum of the ideal curve is at 0 dB. The maximal power of the target is decreasing with increasing maximal position error.

The main influence on the radar image is the loss of the power of the symmetrical and the unsymmetrical bend of the array plane. In general the signal to noise ratio of the resulting image is decreasing. The effect in a SAR image is much stronger in the case of unsymmetrical bend. For the simulation with unsymmetrical bend also the main lobe 3 dB beam width increases slightly and the PSLR degrades. This results in radar image reduced resolution because of the beam widening. The decreased PSLR may result in stronger masking of a weaker target by nearby strong target's side lobes. Results for both position error cases are presented in Table 4.3 and Table 4.4.

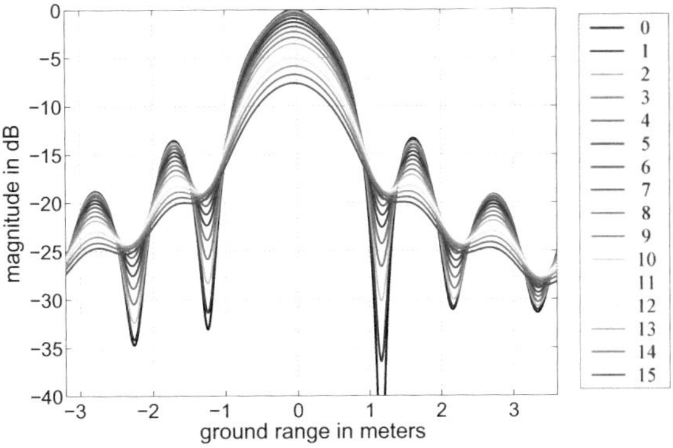

Figure 4.12: Simulation results for SAR with the DBF in range with unsymmetrical position errors.

4.4.3 Comparison of Simulations for Antenna Array and SAR System

The influence of minor position errors on the antenna array characteristic and on the SAR system performance are presented in Section 4.4.1 and Section 4.4.2. The antenna array and the SAR system performance were simulated for symmetrical and unsymmetrical bends. The influence of the symmetrical minor position error on the antenna array is the decrease of the maximal antenna gain at looking direction and the widening of the main lobe. The antenna array with the unsymmetrical bend shows much less decrease of the maximal antenna gain. The main lobe direction is shifted for this type of the antenna distortion. The maximum of the main lobe is for the ideal antenna array simulation at $-1°$ and for the maximal position error of 11 mm the maximum of the main lobe is at about $0°$. Although the drop of the maximum in the main lobe is not so high if considered is the drop at $-1°$ it is much higher then in the symmetrical case.

The simulation for the SAR system is presented in Section 4.4.2. The antenna configuration used for these simulations is the same as for simulations of the antenna array. The symmetrical and unsymmetrical bend has been simulated. Comparing results for all simulations it can be clearly seen that the unsymmetrical bend has much stronger influence on the radar performance.

dx_{max} in mm	Gain loss in dB	3 dB beam width in meter	PSLR in dB
0	0	1.00	13.54
1	0.02	1.00	13.54
2	0.07	1.00	13.52
3	0.15	1.00	13.51
4	0.27	1.00	13.49
5	0.41	1.00	13.48
6	0.60	1.00	13.46
7	0.82	1.00	13.43
8	1.07	1.00	13.41
9	1.36	1.00	13.38
10	1.68	1.00	13.33
11	2.04	1.00	13.29
12	2.43	1.00	13.25
13	2.86	1.00	13.18
14	3.33	1.00	13.10
15	3.84	1.00	13.02

Table 4.3: Results for symmetrical HRWS SAR antenna array bend (AF)

It should be noted here that the considered radar system uses the scan-on-receive DBF. In this technique a narrow beam is 'slided' in range. In this way the received signal is focused by the maximum of the main lobe and by maximum received power for each target. If the antenna beam is disturbed and the main lobe of the antenna array is shifted a performance loss is observed. The angle, by which the main lobe maximum is shifted, can be easily recalculated into ground distance with the assumption that the SAR system is mounted on a satellite platform flying on an orbit at about 780 km height over ground with an incidence angle of $30°$. A one degree shift in the direction of the radar

dx_{max} in mm	Gain loss in dB	3 dB beam width in meter	PSLR in dB
0	0	1.00	13.54
1	0.05	1.00	13.56
2	0.17	1.00	13.52
3	0.36	1.01	13.70
4	0.61	1.01	13.76
5	0.94	1.02	13.80
6	1.34	1.03	13.81
7	1.79	1.04	13.81
8	2.32	1.05	13.76
9	2.91	1.06	13.68
10	3.57	1.07	13.57
11	4.29	1.09	13.41
12	5.06	1.10	13.23
13	5.88	1.12	12.99
14	6.74	1.13	12.72
15	7.62	1.14	12.39

Table 4.4: Results for unsymmetrical HRWS SAR antenna array bend (AF)

beam results in about 18 km shift of the main beam position on the Earth's surface. For the unsymmetrical bend the performance loss can be also seen in resolution and PSLR values.

The difference between the power loss at -1° by the antenna array simulation and the power loss in the SAR simulation can be explained with a specific time extension of the radar signal. The simulated radar signal has the duration of 70 μs. This means that the receive signal extends over in this case about 0.5° of the antenna characteristic. This effect can be, if necessary, easily compensated in SAR processing. The influence gets stronger with narrower beam. The unsymmetrical bend is more degrading on resulting SAR performance. It has much stronger influence on the power loss, the resolution of the system and the side lobes suppression of the target.

4.5 Summary

The influence of minor position errors on the performance of the antenna array with the DBF in range and also on the performance of the SAR system with such antenna configuration was presented in this chapter. A single HRWS SAR receive array panel was used as a model for the simulation of the radar system. First the influence of a single element misplacement has been described analytically for a general case. The array factor for an ideal antenna array and the same array with position errors has been calculated. The array factor for a 60–element array with the symmetrical and unsymmetrical minor position error was calculated. The influence of those two types of errors was simulated for an antenna array. These minor position errors were also introduced to the HRWS SAR with the scan-on-receive DBF in range. The simulation used in previous chapter was adapted for this purpose.
It has been shown that the symmetrical minor position error has more influence on the maximum of the main beam of the antenna array. The unsymmetrical minor position error in antenna array results in a shift of the main receive lobe. This has a strong influence on the performance of the SAR system with the the scan-on-receive. The unsymmetrical minor position error results in power loss, resolution and PSLR decrease. This type of position error degrades the performance of the HRWS SAR system much stronger than the symmetrical bend.

5 Low Bit Quantization

The quantization of signals, especially low bit quantization in multi-channel radar systems is considered in this chapter. Radar signals received by an antenna are continuous in time and amplitude. Such signals are described as analog signals. The processing of analog signals requires dedicated hardware components. The analog processing system once designed and built cannot be changed or reprogrammed. It cannot be adjusted or modified later. If a change in the processing algorithm is necessary the system often must be redesigned and built from scratch, which is again a time consuming and expensive process. Digital signals which are discrete in time and amplitude are much easier to handle[1], [51]. Digitized signals can be easily stored and processed later. This feature is important for a satellite system, where the data is collected, compressed and stored to be transmitted to a ground station when the satellite is flying over. Most of the radar signal processing takes place after transmission of the data to a ground station. A digital processing system can be, to a large extent, adjusted without the necessity of redesigning it. A change in the processing algorithm can usually be applied with a software update. This is advantageous especially for e.g. satellite systems, where a hardware update is not possible.

The amount of recorded data for a satellite radar system is very high. All recorded data must be transmitted to a ground station in a time window while satellite is flying over it. The data are usually compressed before transmitting with an BAQ algorithm [98], [5]. The HRWS SAR records data of multiple independent receive channels. This gives a high amount of recorded data to be transmitted. Transferring received and stored raw radar data from for example satellite to a ground station is expensive and time consuming. One possibility

[1] The expression digital signal and signal in discrete time do not have exactly the same meaning. In both cases a signal is considered, where time has been discretized. For the digital signal also values of signal amplitude are quantized. This must not necessarily be the case for a signal in discrete time, where values of the signal amplitude may be continuous, [51]. In this work an expression quantized signal is used to refer to a digital signal, where both time and amplitude values are quantized.

to reduce further the amount of the data is to use a low bit quantization. The single bit synthetic aperture radar can be found in literature [23], [52], [68], [69]. The novel idea is to use multiple channels and digital beam forming algorithms to reconstruct the quantized signal. A low number of quantization bits would significantly reduce the amount of data to be transferred. It is although to be taken into consideration that a too low number of quantization bits can implicate ambiguities in the SAR image.
The general principle of quantization the process is described in Section 5.1. First considered is the quantization of signal with a low input noise level like for e.g. music. Quantization of those signals with a relatively low input noise level has been well described in literature [95], [51], [58], [73]. Useful and simple rules of thumb are available for such signals, which make it easy to estimate necessary quantization parameters for a specified digital signal quality. Then the quantization of signals with a high level of noise is described. Estimations used for the quantization of signals with a low input SNR are not adequate in this case. The model of a quantized signal with high noise must consider not only the analog and resulting digital signal but also the analog input noise and its influence on the quantized signal. The focus of this chapter is low bit quantization in a multi-channel radar system, where the input signal is quantized with only few bits. The HRWS SAR demonstrator has been measured as an example of such system. Results are presented in Section 5.2. This measurement shows the principle of the multi-channel low bit quantization in praxis. Simulations have been conducted to better understand the dependence of the low bit quantization on number of independent channels, input SNR_i and number of quantization bits. Results of all simulations are presented in Section 5.3. A multi target scenario is considered in Section 5.4.

5.1 Quantization and Sampling

The amplitude and time of an analog signal are both continuous. Quantization of an analog signal is demonstrated in the following example. Analyzed is a real valued chirp signal as presented in Fig. 5.1.
An ideal ADC (analog-to-digital converter) converts the input chirp signal, which is continuous in time and in amplitude, into a discrete in time as well as in amplitude representation – digital signal, [58]. The physical realization of the ADC can be seen as two components: a sample-and-hold block and

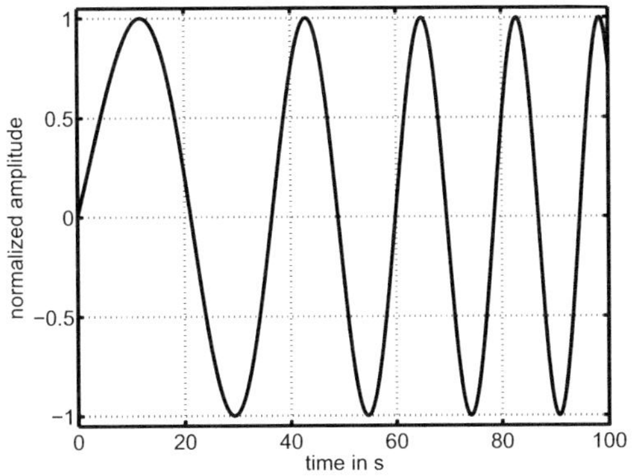

Figure 5.1: Example of an analog signal - a chirp signal.

a digitization of amplitude. First the sample-and-hold block, as the name indicates, holds the input signal amplitude for the time T_s, where T_s is the time duration between two following samples. The output of this component can be described by Eq. (5.1).

$$x_s = \sum_{n=-\infty}^{\infty} x[n] h_0(t - nT_s) = h_0(t) * \sum_{n=-\infty}^{\infty} x_A(nT_s)\delta(t - nT_s) \quad (5.1)$$

Where $x_A(nT_s)$ is the analog input signal for the time nT_s, $\delta(t)$ is the Kronecker delta and $x[n]$ is the quantized output signal with continuous amplitude values. $h_0(t)$ is the impulse response of the zero-order-hold system, defined as in Eq. (5.2).

$$h_0(t) = \begin{cases} 1 & 0 < t < T_s \\ 0 & \text{otherwise} \end{cases} \quad (5.2)$$

The output signal from the sample-and-hold component is provided to the input of the actual A/D converter, which digitizes amplitude. The amplitude is represented with the help of a binary code. The value of the output amplitude is the closest digital value[2] to the analog amplitude of the input signal. The precision of this conversion depends on the number of quantization bits b and

[2] The exact value might differ depending on the implementation of the ADC and the metric used to compare the analog and digital value.

on the dynamic range X_m of the ADC with relation to the maximal amplitude of the input signal. Eq. (5.3) shows the dependence between the number of amplitude levels N_m and the number of bits b. The sample-and-hold block and the A/D converter, are controlled by an external clock. For each time interval T_s a sample of the digital signal is provided.

$$N_m = 2^b \tag{5.3}$$

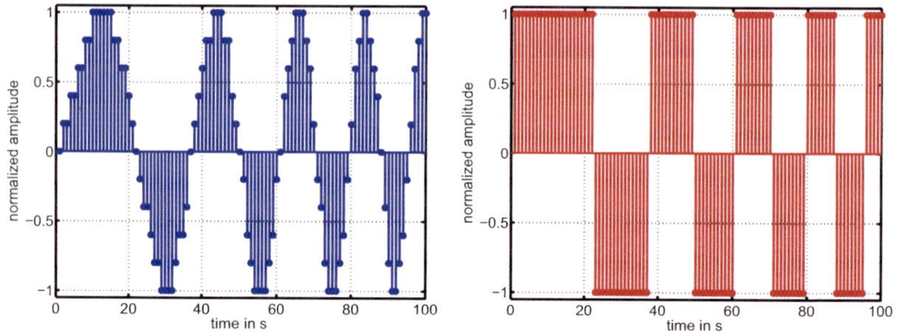

(a) Multiple bit quantization.

(b) One bit quantization

Figure 5.2: Quantization of a chirp signal.

Fig. 5.2(a) shows the input chirp signal quantized with 11 amplitude levels. The value of 11 was chosen to make the steps in images clearly visible. Fig. 5.2(b) shows the same signal quantized with only one bit that gives two amplitude levels often referred to as low and high. Fig. 5.2(a) and Fig. 5.2(b) show digital signals with digital amplitude and time resulting from the same input signal quantized with 11 and 2 quantization levels accordingly. The amplitude of the signal is sampled with a fixed sampling frequency $f_s = \frac{1}{T_s}$. The digitization of time is indicated in figures by using a stemp plot. The dependence between the analog signal at times $t_i = [0, T_s, 2T_s, ...]$ can be expressed with Eq. (5.4). $x_A[t_i]$ is the analog signal at times t_i, $\hat{x}[t_i]$ is the digital signal and $n_q[t_i]$ is the quantization error also referred to as quantization noise.

$$x_A[t_i] = \hat{x}[t_i] + n_q[t_i] \tag{5.4}$$

Fig. 5.3(a) shows the quantization error $n_q[t_i]$ for the example chirp signal quantized with 11 amplitude levels. In the example the parameters of the

ADC have been adjusted to the input analog signal, with sufficient number of amplitude levels (quantization bit) and optimal usage of the full dynamic range of the ADC. It should be noted that the quantization noise is not correlated with the input signal.

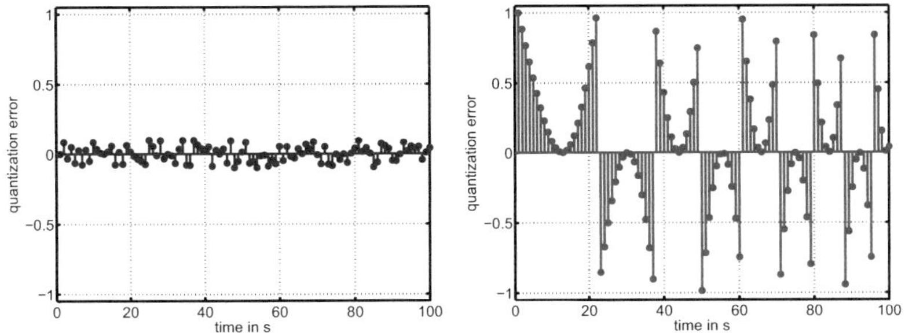

(a) Multiple bit quantization. (b) Single bit quantization.

Figure 5.3: Quantization error for a digital signal.

In the general case assuming that parameters of the ADC have been properly adjusted to the quantized analog signal (high enough number of the quantization bits is used and the dynamic range is exploited), following statistical properties of the quantization noise can be observed, [58]:

- the quantization error is a sequence of a stationary random process
- the quantization error $n_q[t_i]$ is uncorrelated with the signal sequence $\hat{x}[t_i]$
- the quantization error is a white noise process, which means the random variables of the error process are uncorrelated
- the probability distribution of the error process is uniform over the range of the quantization error

These properties concern the quantization noise of signals with no or low additive input noise quantized with a high number of bits. Low bit quantization is a case which does not necessarily obey the same rules. Fig. 5.3(b) shows the quantization error for the example signal quantized with only one bit. It can be clearly seen that the quantization noise is strongly correlated with the original analog signal. For quantization with a low number of bits assumptions presented above for the quantization error are not valid. In this case the

quantization noise is more correlated with the analog signal than the resulting digital signal. The quantization error cannot be assumed a stationary random process and properties of the quantization error depend strongly on properties of the analog input signal.
The HRWS SAR has multiple independent receive channels. All channels in elevation direction are combined with a DBF algorithm. The quantization properties of signals with low noise level are described in following Section 5.1.1. In Section 5.1.2 an input signal with high input noise is considered. The optimal low bit quantization of multi-channel system is considered in Section 5.1.3.

5.1.1 Quantization of Signals with Low Input Noise Level

In this section the quantization of an analog signal with no or low level of an analog additive input noise is described. If the noise level is significantly lower than the signal level it can be considered in the quantization model as noiseless signal. The sampling of a music track is a good example of such a signal, where the input noise level at the ADC is negligible. Eq. (5.4) shows the dependence between the analog and the digital signal.
The quantization noise can take values from a specific interval as described in Eq. (5.5). The maximal quantization error is equal to half the quantization step Δ.

$$-\frac{\Delta}{2} < n_q[t_i] \leq \frac{\Delta}{2} \tag{5.5}$$

The number of quantization bits b, the quantization step Δ and the dynamic range X_m are related to each other according to Eq. (5.6). The dynamic range X_m defines the maximal amplitude of the quantized signal. A signal with a maximal amplitude above this value will reach saturation in the ADC. Quantization parameters should be chosen such that the dynamic range is optimally used. Then the system does not go into saturation and all amplitude levels are used.

$$\Delta = \frac{X_m}{2^b} \tag{5.6}$$

The figure of merit for the quality of the digitized output signal is the signal-to-noise ratio SNR_q for wanted input signal and the quantization noise. SNR_q

is defined as a ratio of the signal power to the noise power. The index $_q$ indicates that the ratio is calculated against the quantization noise.

$$SNR_q[dB] \quad = \quad 10\log_{10}\frac{\sigma^2_{signal}}{\sigma^2_{quantization\ noise}} \tag{5.7}$$

SNR_q can be easily estimated for a sinusoidal signal, [58]. The general rule of thumb is that SNR_q increases by about 6 dB with every additional quantization bit. The exact SNR_q calculation is given in Eq. (5.8).

$$SNR_q[dB] = 6.02b + 10.8 - 20\log_{10}\frac{X_m}{\sigma_x} \tag{5.8}$$

σ_x is the RMS (Root Mean Square) value of the signal. For a sinusoidal signal with a maximal peak amplitude of X_p, σ_x is equal to $\frac{X_p}{\sqrt{2}}$. It must be remembered that Eq. (5.8) does not consider possible saturation effect. This happens when σ_x is too high ($X_p \gg X_m$). If σ_x is too small, the signal-to-noise ratio in Eq. (5.8) decreases. It is therefore to be assured that the signal amplitude and the full-scale amplitude dynamic range of the ADC are carefully matched.

In general, a signal with high input SNR_i can be described with Eq. (5.9). The signal $x(t)$ consists of the wanted signal $x_i(t)$ and the input noise $n_i(t)$. For a signal with low input noise it can be assumed that the input signal is equal to the wanted signal.

$$x(t) = x_i(t) + n_i(t) \approx x_i(t) \tag{5.9}$$

Another figure of merit for the quantized signal is the Mean Square Error (MSE). Here it is assumed that $n_i(t)$ is negligible compared to the input signal $x_i(t)$, as in Eq. (5.9). The input signal is quantized with a relatively high number of bits and the fitting of the dynamic range is ensured for the signal. The MSE can be calculated as the difference between the quantized signal $\hat{x}[t_i]$ and the original input signal $x[t_i]$ for times t_i. MSE is calculated according to Eq. (5.10).

$$MSE = \left\langle \left|x[t_i] - \hat{x}[t+i]\right|^2 \right\rangle \tag{5.10}$$

5.1.2 Signals with High Level of Noise

The radar receive signal is usually quantized with 8 to 16 bits depending on system parameters and requirements. This quantization is sufficient to extract

the information enclosed in the receive signal. Radar signals, compared to other digital signals like e.g. music, have quite low input signal to noise ratio. Often the noise level is much higher than the wanted radar receive signal. The information enclosed in the signal is extracted through adequate signal processing.
In the case of a signal with a very high input signal–to–noise ratio or even without input noise, the whole input signal is considered as the wanted signal. An analog signal with high amount of input additive noise is considered now, where the input noise cannot be neglected. In this case the 'traditional' calculation for the quantization output SNR_q according to Eq. (5.8) as well as the calculation of the number of necessary bits for optimal quantization is no longer valid. If the input signal is quantized with parameters calculated for a noiseless signal the additional bits are spent rather on noise and not to describe the wanted signal. A description of the one bit and low bit quantization can be also be found in [89].
In the ADC model for signals with low input SNR_i the noise must also be considered as a part of the analog input signal. The figure of merit used for the input signal with low noise level is not applicable for signals with high noise level. When the noise power is higher than the wanted signal power (input SNR_i in dB is negative) the MSE criterion suggests that the minimum quantization error is smaller for a lower number of quantization bits. The detailed description of the calculation of MSE for the high noise case can be found in [79]. A new figure of merit has been used for the quantization of signals with high noise level in [79]. The new metric includes a gain factor g_f which scales the input signal. The gain factor is chosen so that the estimated noise power is minimal, Eq. (5.11). The resulting output SNR_o, combining the quantization noise and input noise, can be calculated according to Eq. (5.12). With this correction the estimation is valid also for signals with a high content of noise on the input quantized with a low number of bits.

$$g_f = \frac{< x_o s_i >}{< s_i^2 >} \tag{5.11}$$

$$SNR_o = \frac{< s_o^2 >}{< n_o^2 >} = \frac{< s_o^2 >}{< \frac{1}{g_f} - s_i >^2} \tag{5.12}$$

It is important for an optimal quantization of analog signals to match the saturation amplitude X_m of the ADC with the maximal amplitude of the analog signal. In following section an estimation of the optimal saturation amplitude for the ADC for low SNR_i input signals quantized with a low number of bits is described.

5.1.3 Optimal Saturation Amplitude of the Low Bit ADC

In most cases an amplifier is introduced before the ADC to ensure optimal matching of the signal power level. A part of the normalization can also be done by a specially designed filter. X_m is the maximal input amplitude of a ADC. If the amplitude X_m is chosen too low a signal with amplitudes higher than X_m will be clipped to the amplitude value X_m. This error of the quantization process is called saturation noise. Otherwise, if X_m is chosen too high the quantized signal loses much of its dynamic. In this case the quantized signal is described only with lower bits, which reduces the resolution of the quantization – altogether less bits are used to describe the signal than in the optimal case.
It should be considered that in the estimation of the optimal saturation amplitude for the input signal with high noise not the whole analog input signal is the wanted signal. Assuming the same values as in the previous case quantization bits are wasted for the noise where the wanted signal is not using the full dynamic range of the quantization process. A solution to this problem is presented in [79]. The optimal saturation amplitude X_{opt} is described with Eq. (5.13).

$$X_{opt} = X_p \sqrt{\sigma^2_{signal} + \sigma^2_{noise}} \tag{5.13}$$

Where σ^2_{signal} is the energy of the signal and σ^2_{noise} is the energy of the input noise.

5.1.4 Systems with Multiple Receive Channels

Systems with a single input channel have been considered in previous sections. The HRWS SAR system improves the performance with a DBF technique. Signals from independent receive channels are combined through adequate superposition of all receive channels in elevation. In this section a radar system with multiple independent receive channels is described. The distinctive feature of the HRWS SAR system are multiple independent receive

channels with a high additive input noise. The signal after the DBF processing can be seen as a superposition of N independent channels each sampled with N_m amplitude levels.
The maximal possible number of amplitude levels of the output after superposition of real valued signals is described by Eq. (5.14)

$$L_{NN_m} = N \cdot (N_m - 1) + 1 \tag{5.14}$$

Where N is the number of independent receive channels and N_m is the number of quantization amplitude levels for each channel. With e.g. $N_m = 2$, which is one quantization bit, the number of amplitude levels is as in Eq. (5.15).

$$L_{N2} = N + 1 \tag{5.15}$$

In case of the HRWS SAR each of the channels is described with the I and Q–component. Both I and Q components in the HRWS SAR demonstrator are digitalized separately, as it is described in Chapter 3. Each of receive signals can be represented as complex signal $s = s_I + j \cdot s_Q$. The DBF in general adjusts the phase of receive signals and builds a sum over all receive signals. The phase difference between receive signals increases with increasing signal reception angle. Receive signals from targets positioned in front of the antenna array arrive all at the same time, the difference phase angle is small. Receive signals from target positioned sideways arrive with time delay depending on the distance difference between target and antenna array elements. The number of possible amplitude levels the signal after the DBF algorithm can take is in this case unlimited. This is because of the phase adjustment for a complex receive signal. The one bit quantized complex signal is multiplied with a complex value containing phase shift.
The idea of low bit quantization is illustrated in Fig. 5.4. Five independent receive channels are indicated in the figure. For the purpose of demonstration used are real valued signals. It should be noted that each of received signals has a different initial phase. This can be seen in the figure by looking at the middle of the signal. The phase difference between received signals is calculated for a linear antenna array receiving signals coming from directions not perpendicular to the antenna array plane. The analog receive signal is plotted for each of five channels (upper plots). Analog signals are quantized with one bit. Middle plots show one bit quantized input signals. Each of those signals has now only two amplitude levels – low and high. Next the

DBF algorithm is applied. Adequate phase shift is applied to each of those signals and the sum is built. The signal resulting after the DBF algorithm is presented in the lower plot. All five one bit quantized signals are combined in it. The original chirp signal is reconstructed.

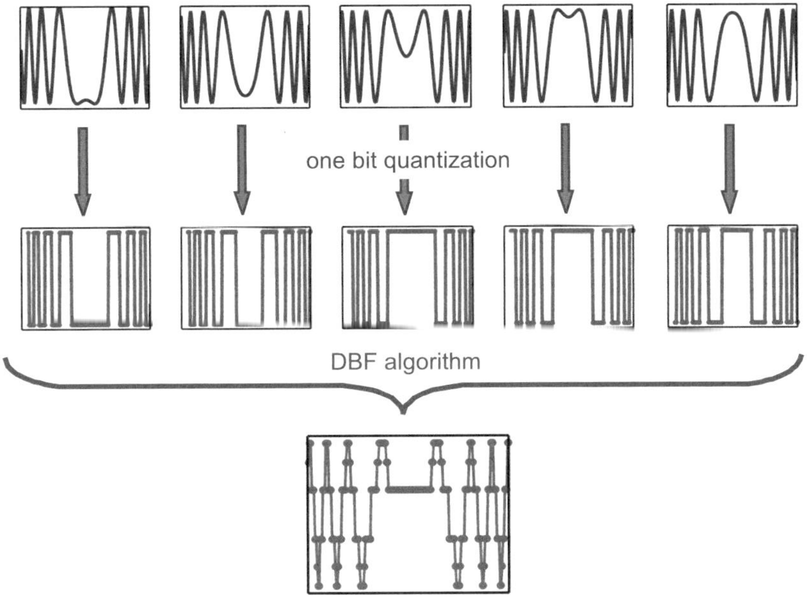

Figure 5.4: Illustration of low bit quantization and DBF.

In systems with multiple independent receive channels the amount of data increases significantly with each additional receive channel. Reducing the number of quantization bits would reduce the complexity of the system, cut costs and reduce amount of the raw unprocessed data. The idea of low bit quantization in radar systems leans on several assumptions which are:

- the analog signal contains a high level of additive white noise
- the system has multiple independent receive channels
- the system uses a DBF algorithm to reconstruct receive signal

The HRWS SAR demonstrator is an example of a multi-channel radar system with one bit quantization. In following section the measurement of the demonstrator and reconstruction of original signal upon real measured data is presented.

5.2 Measurements with One Bit Quantization

The principle of one bit quantization in a multiple channel system was described in the previous section. A HRWS SAR demonstrator has been measured to demonstrate the principle of the algorithm in practice. The demonstrator consists of eight independent receive channels. On the output of each of the channel provided is a complex IQ signal. Both signals, I and Q, are quantized with one bit. A chirp signal from signal generator was put on the input of the HRWS SAR demonstrator. Additional noise was generated with a noise generator and added to the input signal. A single channel output was measured. To prove the principle of the one bit quantization three different signals were measured: generated chirps signal, generated noise signal and chirp signal with the additional noise. The SNR of the signal was set to -4.5 dB. 66 receive signals were recorded for each of the signal configurations. The coherent sum of the 66 receive signals simulates the effect of the DBF. All receive signals were recorded on the same channel so the phase difference between those signals is zero. Results of measurements can be seen in Fig. 5.5. The amplitude of signals was plotted with values between zero 'low' and one 'high'. Signals after the DBF were normalized to the maximum value of one. Left side plots show a single receive signal for the chirp signal, the noise and the chirp signal with noise. Right side figures show a corresponding superposition of 66 receive signals. Fig. 5.5(a) shows the single receive chirp signal. The chirp is clearly visible. The number of amplitude levels is very limited. Fig. 5.5(b) shows the superposition of 66 received chirp signals. No significant difference can be seen between the single chirp and the superposition of 66 receive signals. The chirp signal has not been reconstructed through the DBF. Fig. 5.5(c) shows the received noise signal and Fig. 5.5(d) show a superposition of 66 of those signals. A suppression of noise in Fig. 5.5(d) can be observed. The number of various amplitude levels is also significantly higher. The suppression of noise depends on the number of superimposed receive signals. The expected value is 0.5, which corresponds with the achieved result. Fig. 5.5(e) shows a single receive chirp signal with additional noise quantized with a single bit. The receive chirp is not visible in this figure because of the additional noise. Fig. 5.5(f) shows the superposition of 66 received chirps signal with additional noise. In this figure the chirp is clearly visible. To notice is also the much higher number of amplitude levels. The chirp signal has been reconstructed through an adequate DBF processing.

The principle of the one bit quantization and reconstruction of the signal with the DBF can be explained by comparison of plots on the right side in Fig. 5.5. The chirp signal without noise was not reconstructed. The number of amplitude levels is more or less the same as before the DBF. It is because receive signals are very similar to each other. There is no phase shift between them as they all were measured on the same receive channel. The one bit quantized noise signal is a random process and has different values for each of measurements. The expected value for a white noise is equal to its mean value, in this case 0.5. This can be observed in Fig. 5.5(d). The values of the signal after the DBF approach the expected value. The result improves on quality with the increasing number of receive signals incorporated into superposition. The chirp signal with additional noise can be reconstructed by the means of the DBF. This can be observed in the final plot. The additional noise is suppressed and the received chirp signal can be clearly seen. This shows that additional noise is an important component for reconstruction of the original signal through DBF algorithm.

The measurement shows one example of a multi receive channel system. It is difficult to built a hardware for each parameter variation of the system. In following section described are results of a HRWS SAR simulation. Optimal quantization parameters are found with help of the simulation. Simulates is the dependence on the input SNR_i of a single channel, the number of independent channels and the number of quantization bits.

5.3 Simulations of Optimal Quantization

Various factors must be taken into consideration in order to parametrize a system with multiple receive channels. Parameters: number of receive channels, number of quantization bits and input SNR_i for optimal quantization of the input signals in multiple receive systems with the DBF are described in this section. SNR_i is here the signal-to-noise ratio of the input signal for each independent channel. It is assumed that all receive channels are identical

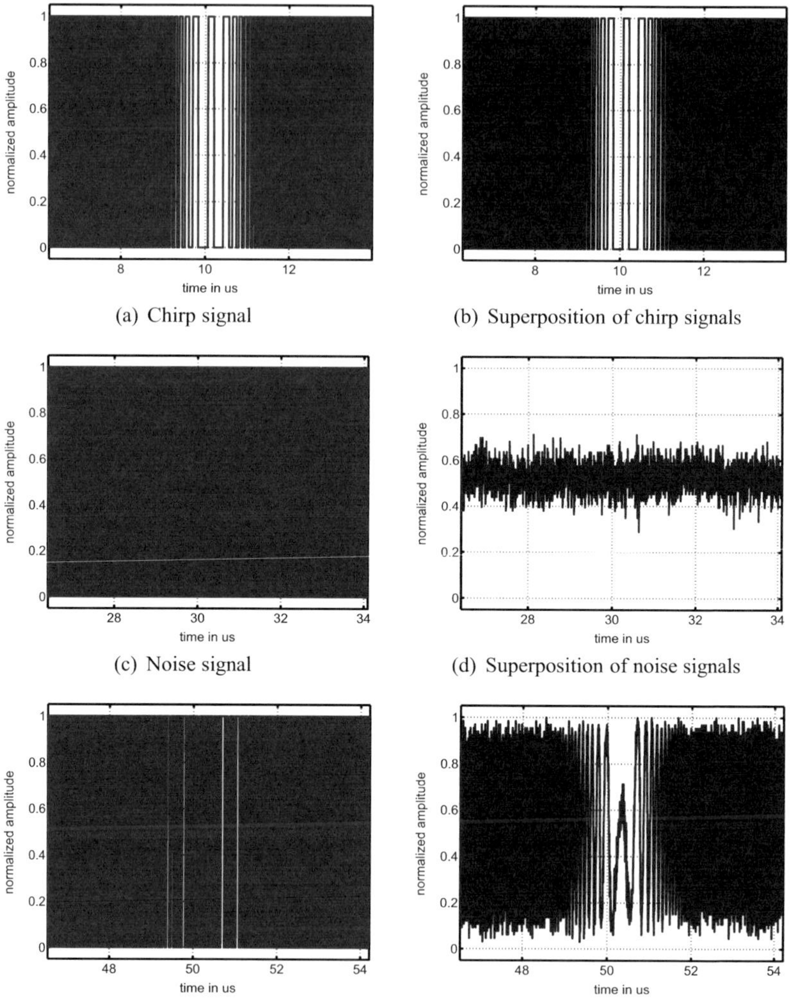

(a) Chirp signal

(b) Superposition of chirp signals

(c) Noise signal

(d) Superposition of noise signals

(e) Sum of the noise and chirp signal

(f) Superposition of noise and chirp signals

Figure 5.5: Single chirp, noise and chirp plus noise signals and a coherent sum of 66 pulses. The SNR_i equals -4.5 dB in terms of the hardware design and hardware components as well as input SNR_i. SNR_o is the signal-to-noise ratio of the output signal of the SAR system after quantization, the DBF algorithm and the application of a SAR processing on the receive radar data. In this case the processing gain is also considered. SNR_o is estimated as the noise floor level – a difference between the average noise floor level and the maximal signal amplitude after compression. For each simulation the same transmit radar signal with specified bandwidth and time duration has been used and also the same general SAR processing has been applied. Special interest is put on systems with low input SNR_i on the single input channel, as this is usually the case in spaceborne radar systems. Following sections provide radar system simulation results describing the dependence of SNR_o on input SNR_i on each channel, the number of independent receive channels N and the number of quantization bits b. For each case regarding various SNR_i and number of receive channels also one bit quantization is considered.

5.3.1 Dependence between the SNR_o and the Number of Receive Channels

The dependence between the number of independent receive channels and the output SNR_o is investigated in this section. Additional information that allows for improvement of the resulting radar image is gained with each additional receive channel. Simulations have been performed for the number of independent receive channels N equal to 1, 2, 3, 5, 10, 20, 40, 60, 120 and 240. The amplitude of resulting signals has been normalized so that results can be easily compared with other simulations. The assumption is that every receive antenna sub-aperture has the same signal parameters at its input. This means the same input SNR_i and the same signal power. For this simulation SNR_i has been set to -10 dB. The radar signal used in the simulation had the bandwidth B of 300 MHz and the time duration of the transmitted chirp τ of 70 μs. The antenna array was a uniformly spaced array with a spacing h_{rs} of 16.7 cm. A single receive panel consisting of N by 1 sub-apertures was simulated. The chosen scenario consists of a single point target positioned on the Earth's surface.

Fig. 5.6(a) and Fig. 5.6(b) show results of simulations with various numbers of receive channels. Fig. 5.6(a) presents signals without quantization. Fig. 5.6(b) presents results for the case of signals quantized with one bit before the DBF algorithm and the SAR processing are applied. As it can be seen for both cases the general rule is that with increasing number of independent receive channels the SNR_o of the output signal improves. It should be noted that the improvement is quite high for a smaller number of channels, while the change between e.g. 120 and 240 channels is relatively small. The difference between one aperture and three sub-apertures in SNR_o is about 5 dB. The difference between 120 and 240 sub-apertures in SNR_o is about 3 dB. The SNR_i improves by 3 dB with doubling of the number of receive sub-apertures. Another observation is that for each corresponding simulation with the same number of independent receive channels the one bit signal has lower SNR_o than the signal with no quantization applied.

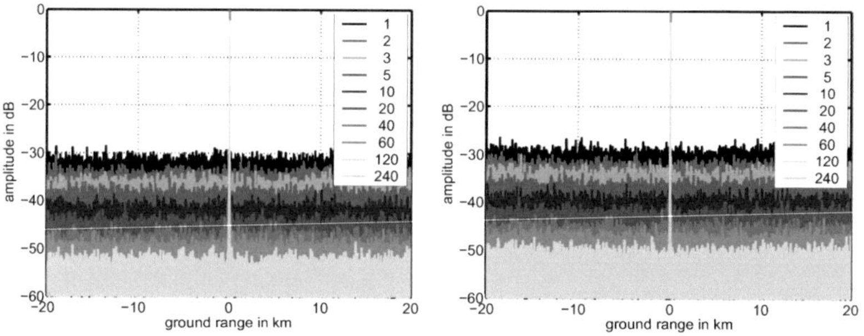

(a) No quantization

(b) One bit quantization

Figure 5.6: Simulation for various values of receive channels N

The values for SNR_o for all simulations presented in Fig. 5.8 can be found in Table 5.1. Fig. 5.7 presents output signal-to-noise SNR_o values in dependence on the number of sub-apertures N in the receive antenna. The red curve corresponds to simulation results without quantization and the blue curve to simulations with one bit quantization. Both curves look very similar in shape. It is to be noted that there is a constant difference between the curve for no quantization and the curve for one bit quantization. This difference is about 2 dB. The simulation showed it will not be eliminated by the increasing number of independent receive channels. It remains constant with increasing number of receive channels.

N	noise floor no quantization in dB	noise floor for 1 bit in dB
1	-41.4	-39.2
2	-44.5	-42.2
3	-46.2	-44.0
5	-48.4	-46.2
10	-51.4	-49.2
20	-54.4	-52.2
40	-57.4	-55.2
60	-59.2	-57.0
120	-62.2	-60.0
240	-65.2	-63.0

Table 5.1: Output noise floor level in dependence on the number of receive channels N, $SNR_i = -10$ dB.

The dependence between SNR_o and the number of receive elements is logarithmic and can be described with Eq. (5.16).

$$SNR_o = SNR_i + 10\log_{10}(B\tau) + 10\log_{10}(N) \tag{5.16}$$

This shows that the signal-to-noise ratio for a radar system with multiple receive channels improves with increasing number of channels. The improvement is although much more rapid for smaller numbers of receive channels. The change is significant for N up to about 50 sub-apertures. Then until N of about 100 some improvement is still noticeable but it is achieved with high effort. The improvement is 3 dB for each doubling of the number of receive sub-apertures. In practice to use more than about 60 channels would be costly and mechanically challenging without measurable benefit in SNR_o. Above values of N of 100 the additional improvement with each sub-aperture is low. The second observation is that there is a constant difference between the SNR_o of the reference signal and the signal quantized with one bit. This gap

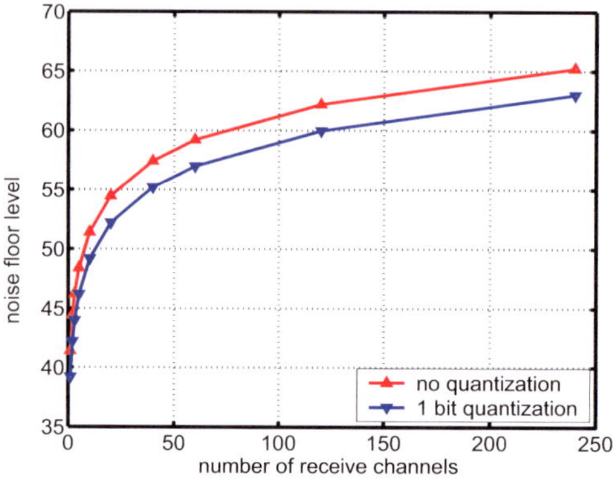

Figure 5.7: Noise floor level in dependence on the number of receive channels N, $SNR_i = -10$ dB.

cannot be reduced with higher number of receive channels. This corresponds to results described in [79].

5.3.2 Dependence between the SNR_o and SNR_i

The dependence between SNR_i of the input signal on each independent receive channel and the output SNR_o of the resulting compressed radar signal is described in this section. Simulation parameters were chosen identical as previously described in Section 5.3.1 with B of 300 MHz, τ of 70 μs and an antenna array element spacing of 16.7 cm. The number of independent receive channels N was set to 20. The point target used in the test scenario is positioned in front of the radar system on the line of sight of the SAR transmit and receive antenna. Simulations have been made for signals without quantization and with one bit quantization on each channel for various values of input SNR_i.

In Fig. 5.8(a) and Fig. 5.8(b) simulated compressed radar receive signals are presented. The simulation was conducted for input SNR_i of -20dB, -15dB, -10 dB, -5 dB, -3 dB, 0 dB, 3 dB, 5 dB, 10 dB, 15 dB and 20 dB for both quantization cases. Simulation results for not quantized signals are shown in Fig. 5.8(a). As expected with increasing SNR_i the SNR_o is also increasing accordingly. In Fig. 5.8(b) results for simulations with one bit quantization on each independent receive channel are presented. Here also the SNR_o increases with improvement of the input SNR_i. The relation between SNR_i and SNR_o for signals with no quantization is linear. This is although not the case for signals that are quantized with one bit. An improvement in SNR_o is not noticeable in Fig. 5.8(a) for SNR_i higher than about 3 dB.

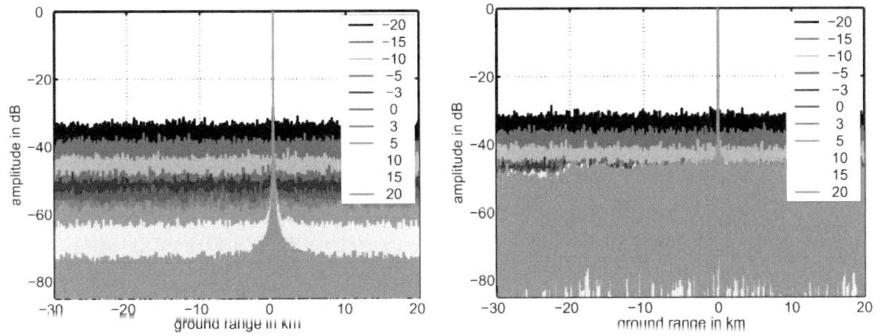

(a) No quantization signal (b) One bit quantization

Figure 5.8: Simulation for various values of SNR_i

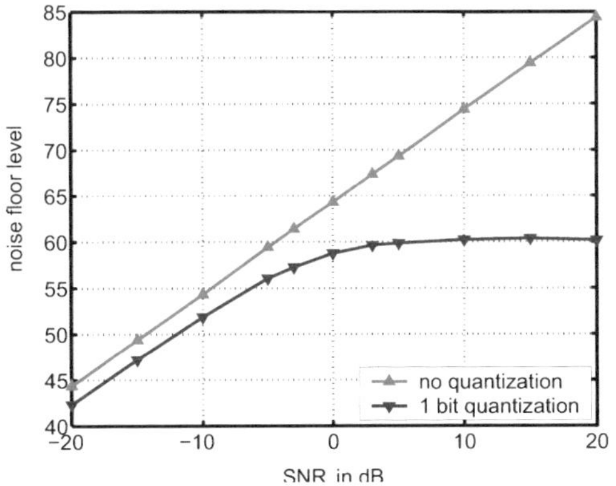

Figure 5.9: Noise floor level in dependence on SNR_i for one bit quantization and not quantized signal, $N = 20$.

Fig. 5.9 presents curves representing the simulated dependence of noise floor level on SNR_i for signals with no quantization (red curve) and quantized with one bit (blue curve). The values of noise floor level for all simulations are presented in Table 5.2. The difference between curves for the one bit quantized signal and the reference signal with no quantization can be clearly seen in Fig. 5.9. The noise floor level for the not quantized signal decreases continuously with decreasing SNR_i. This linear dependence is expected as the SAR system can be considered as linear system. The curve for the one bit quantized signal shows the same behavior for low values of SNR_i – less than about -5 dB. Although for higher values of SNR_i the curve comes to a saturation, and further increase of SNR_i does not improve the SNR_o. The maximum value of noise floor level for one bit quantized signals within these system parameters is about 60 dB. A further increase of the SNR_i does not improve the value for resulting noise level over this value. These results show that although one bit quantization is effective for signals with high additive white noise, where SNR_i is lower than about -5 dB, it is not efficient for signals with low noise level with SNR_i better than about 0 dB. Further improvement of the input signal does not provide a better radar image. One bit quantization is therefore useful only for signals with very low input SNR_i. Analyzing results presented in Fig. 5.9 and Table 5.2 it is to be noted that for SNR_i values between -20 dB and -5 dB there is a constant difference between the noise floor level of the reference signal and the noise floor level of the one bit quantized signal. For low values of SNR_i two curves for the not quantized signal and the signal quantized with one bit are parallel to each other with about 2 dB constant difference between them. Presented results have been acquired for the number of receive channels N equal to 20. A higher number of channels does not reduce the minimal difference between the one bit quantized signal and the reference signal as it was already shown in Section 5.3.1. This shows that although the one bit quantization simulation for low input SNR_i values provides similar results as the simulation without quantization, a 2 dB performance loss must be accounted for.

5.3.3 Dependence between the SNR_o and the Number of Quantization Bits

The dependence of the output SNR_o on the number of quantization bits is described in this section. Simulations have been repeated with the same system parameters for a various number of quantization bits. SAR system parame-

SNR_i in dB	noise floor no quantization in dB	noise floor for 1 bit in dB
-20	-44.3	-42.3
-15	-49.4	-47.2
-10	-54.3	-51.9
-5	-59.4	-56.0
-3	-61.4	-57.3
0	-64.3	-58.8
3	-67.4	-59.6
5	-69.3	-59.9
10	-74.4	-60.2
15	-79.4	-60.4
20	-84.4	-60.2

Table 5.2: Output noise floor level in dependence of the input SNR_i, $N = 20$.

ters have been chosen as in previous sections with B equal to 300 MHz and τ equal to 70 μs. The number of independent receive channels N has been set to 20. Simulations have been conducted for input SNR_i of -10 dB. It is shown in Section 5.3.2 that low bit quantization is applicable only for SNR_i values lower than about 0 dB. Simulations have been conducted for 1 to 8 quantization bits. As a reference also a simulation with no quantization in receive channels had been conducted. Simulation results regarding noise floor level are presented in Table 5.3. In Fig. 5.10 results of simulations for various number of quantization bits are presented. The values of noise floor for simulation with various quantization bits is presented in Fig. 5.11(a). The difference between the result for signals quantized with various numbers of bits and the reference signal, which was not quantized is shown in Fig. 5.11(b). The highest difference is as expected between the one bit quantized signal and the reference signal. With increasing number of quantization bits the difference is decreasing exponentially. For 4 bits or higher the difference can be practically neglected. In previous sections it was shown that the difference between the quantized signal and the signal with no quantization cannot be fully compensated by means of increasing the number of independent receive channels or decrease of quality of input signals. In this section it is shown that a minimum of four quantization bits is necessary to avoid performance loss related with the quantization process.

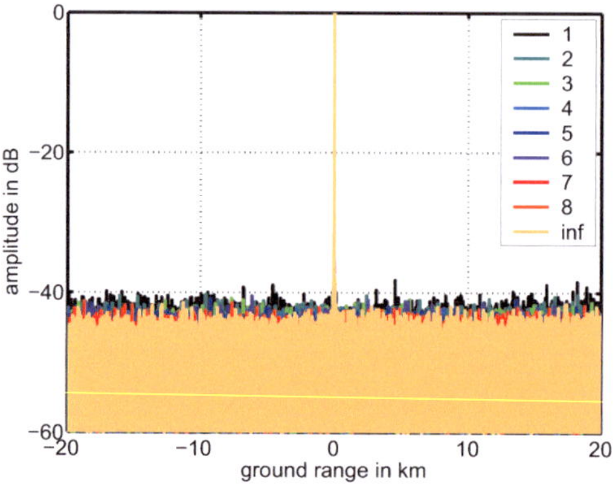

Figure 5.10: Simulation for various values of quantization bit

5.4 Multiple Target Scenario

All previous simulations presented in this chapter considered a single target scenario where a single point target was positioned on the ground. In a multiple target scenario another problem with the low bit quantization for radar signals comes into view. The low bit quantization may result in ambiguities when more than one target is present in illuminated scenario. A real scenario consists of multiple extended targets. A scenario with two closely positioned point targets is considered in this section. The distance between these point targets has been set to about 150 m in range on the ground. Targets are positioned at about -2° of the antenna broad sight. The number of independent receive channels has been set to 60.

Number of bits b	noise floor in dB	difference in dB
1	-52.1	-2.21
2	-53.3	-1.08
3	-54.2	-0.22
4	-54.3	-0.06
5	-54.3	-0.02
6	-54.3	-0.01
7	-54.3	-0.01
8	-54.3	-0.01
no quantization	-54.4	0

Table 5.3: Noise floor level in dependence on the number of quantization bits, for $N = 20$ and $SNR_i = -10$ dB.

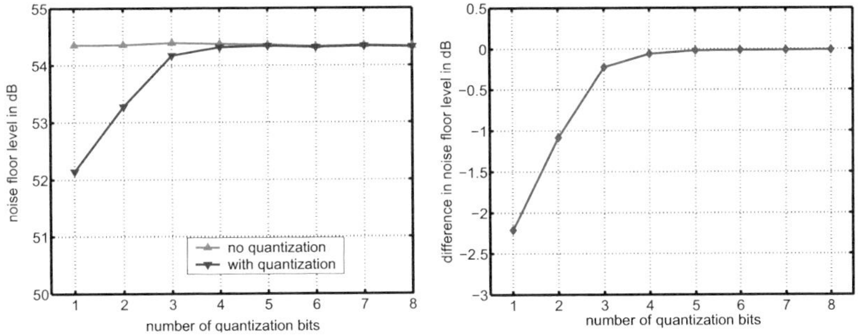

(a) Dependence of the SNR_o on the number of quantization bits

(b) Difference between SNR_o of simulations and the reference for not quantized signal.

Figure 5.11: Simulation for various number of quantization bits. The number of channels was chosen $N = 20$ and $SNR_i = -10$ dB.

Fig. 5.12 from (a) to (f) presents simulation results which differ in the number of quantization bits and/or the value of input SNR_i. Simulated were the case with no quantization, single, two and three quantization bits with various values of SNR_i. Fig. 5.12(a) shows the compressed reference signal with no quantization and SNR_i of 0 dB. Both point targets can be clearly seen in this figure. The output noise level is low. It is to be noted that the noise floor goes down to about the level of -60 dB and no ambiguities are present. Fig. 5.12(b), Fig. 5.12(c) and Fig. 5.12(d) present radar signals simulated with the same system parameters but quantized with one, two and three bits accordingly. In all figures the noise level is significantly higher than in the reference simulation. The difference in the noise floor between Fig. 5.12(a) and Fig. 5.12(b) is about 5 dB. This is consistent with previous results presented in Section 5.3.2. Beside the increased noise floor level an interesting effect can be observed in Fig. 5.12(b) and Fig. 5.12(c). Strong ambiguities occur for signals quantized with one and with two bits. These ambiguities do not occur in Fig. 5.12(d) where receive signals are quantized with three bits. This effect is even more clearly visible by a comparison of Fig. 5.12(b) to Fig. 5.12(e) and Fig. 5.12(f) where signals in all three simulations have been quantized with one bit. Simulations for Fig. 5.12(b), Fig. 5.12(e) and Fig. 5.12(f) were conducted with various values for input SNR_i of 0 dB, 15 dB and -20 dB, accordingly. The dependence between the ambiguities and the input signal SNR_i value can be clearly seen. The strongest ambiguities have about -10 dB to the maximum of the compressed signal occur for the signal with 15 dB input SNR_i quantized with one bit. In Fig. 5.12(b) for the input SNR_i of 0 dB the strongest ambiguities have an amplitude of about -16 dB. It it to be noted that the amplitude of the following ambiguities falls much faster, so that only the first three ambiguities are visible over the noise level and would disturb the resulting image. Similar to the plot for the input SNR_i of 0 dB and one quantization bit is the plot for the same input SNR_i value and two quantization bits. It is to be noted that the first order ambiguities are lower than for the one bit quantized signal and are at about -19 dB. Fig. 5.12(f) shows the resulting radar compressed signal for the input signal with $SNR_i = -20$ dB quantized with one bit. In this case no ambiguities are visible in the figure. The output signal consists of two clear peaks representing two point targets. By comparing Fig. 5.12(f) to Fig. 5.12(b) it can be seen, as expected, that the noise level is higher in the first image. The comparison between Fig. 5.12(a) and Fig. 5.12(f) the difference in the noise level reaches about 22 dB.

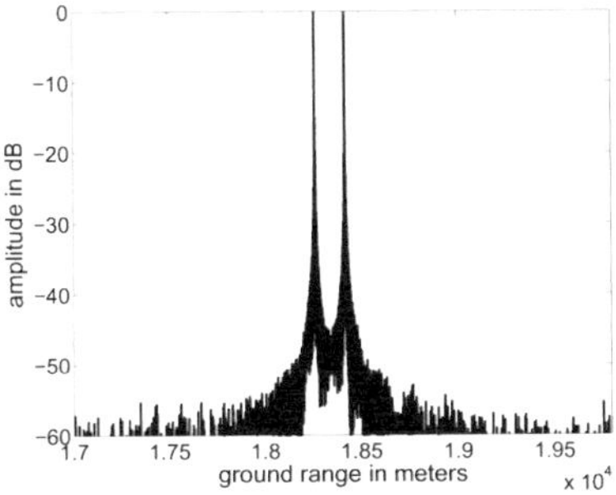

(a) $SNR = 0\ dB$ without quantization.

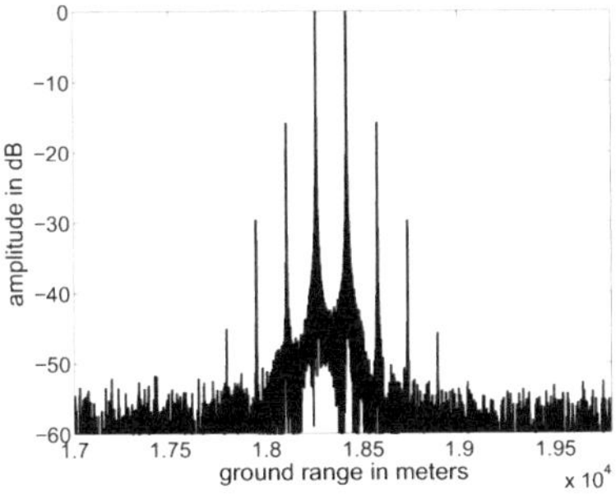

(b) $SNR = 0\ dB$ with 1 bit.

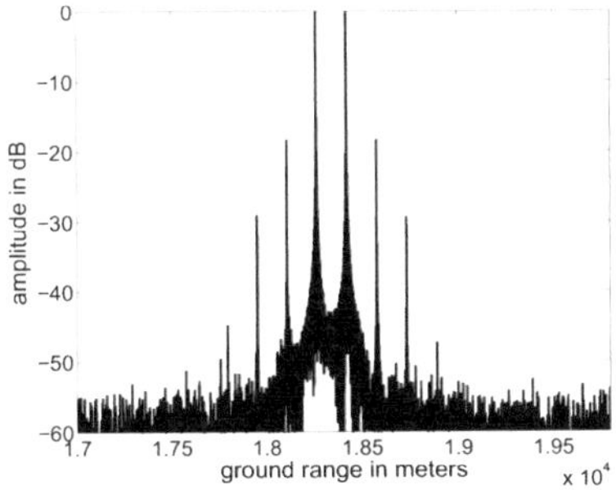

(c) $SNR = 0\ dB$ with 2 bits.

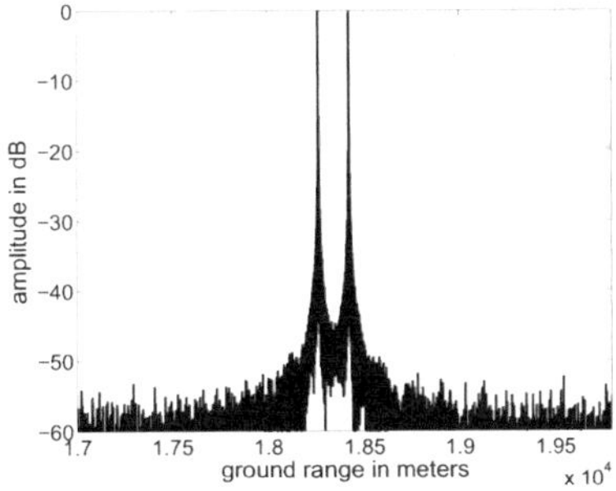

(d) $SNR = 0\ dB$ with 3 bits.

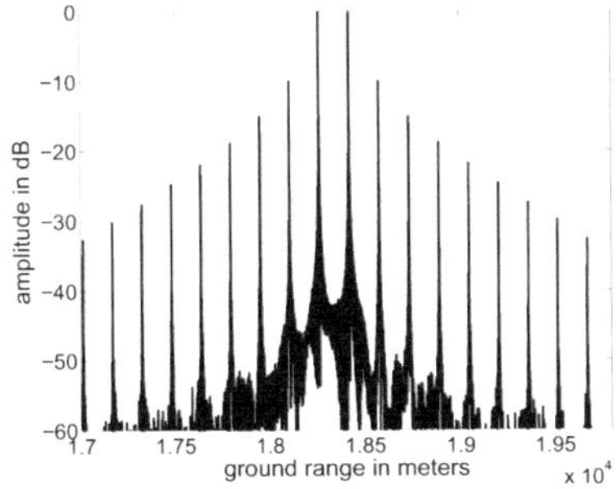

(e) $SNR = 15\ dB$ with 1 bit.

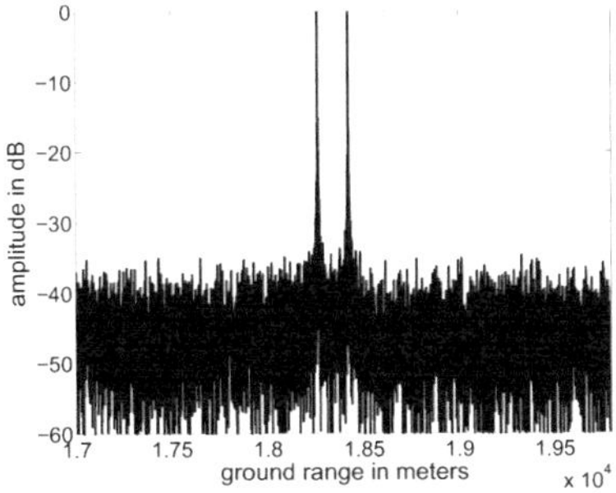

(f) $SNR = -20\ dB$ with 1 bit.

Figure 5.12: Compressed Radar signals with two close positioned targets with various input SNR_i and number of quantization bits.

The possible reason for the occurrence of those ambiguities in radar images quantized with a low number of bits is the quantization noise. In case when many quantization bits are used the quantization noise can be assumed to be evenly distributed. The quantization error is then a white noise process with zero mean and uniform distribution over the whole quantization error range. The most important property in case of a high number of quantization bits is that the quantization noise and the input signal are not correlated. In the case of low bit quantization of signals with low additive input noise, and specially for the one bit quantization, the quantization noise and the input signal are correlated. This can cause ambiguities in the resulting compressed radar image. In the design of a radar system with low bit quantization it must be assured that no ambiguities will occur in the compressed radar image.

5.5 Summary

In this chapter the quantization of a multi–channel radar system with a special consideration of low bit quantization was described. First some basic properties of analog and digital signals were presented. It was shown that general rules and approximations used to parametrize quantization processes are no longer valid for low bit quantization or the quantization of signals with very high additive noise. For those cases the assumption that the quantization noise is a white process and not correlated with the analog input signal, is not fulfilled. In fact in case of low bit quantization the noise is often more correlated with the analog input signal than the resulting digital signal itself.
The principle of reconstruction of one bit quantized signal through the DBF was described and demonstrated. The HRWS SAR demonstrator was measured and the data were processed as an example of multi-channel radar system with one bit quantization. It was shown that a chirp signal without additional white noise cannot be reconstructed through the means of the DBF processing. The same chirp signal with additional white noise was successfully reconstructed.
The optimal quantization parameters were found through the means of the simulation. The dependence of SNR_o on the number of receive channels, the input signal SNR_i and the number of quantization bits was considered. The simulation for various numbers of receive channels and for various SNR_i were conducted for signal without quantization in the receive block

and with one bit quantization. The noise floor level increases with increasing number of independent receive channels, which is consistent with the theory. The simulation showed a 3 dB improvement for doubling of the receive channels. In practice using large number of sub-apertures with independent receive channels would be expensive and mechanically challenging. The improvement for each additional sub-aperture is then insignificant. The relation of the resulting radar image noise floor on the input SNR_i was simulated. Simulations for various input SNR_i of the receive signal showed a relation with the noise floor of resulting radar image. The improvement of the input SNR_i in general improves the resulting SNR_o of the compressed SAR image. This is not the case for one bit quantization. Here the rule is true for values of SNR_i lower than about 0 dB. The resulting SNR_o goes into saturation for values of SNR_i higher than about 5 dB. Further improvement of SNR_i does not improve the resulting SAR image. A constant difference in the noise floor for SNR_i values lower than 0 dB between not quantized and one bit quantized signals was observed. This decrease of SAR performance in signal to noise ratio relates to low bit quantization. It cannot be compensated with increased number of independent receive channels. The loss of about 2 dB for a single bit quantization compared to not quantized signal must be accounted for. This gap depends on the number of quantization bits. It is about 2 dB for a single quantization bit and about 1 dB for two quantization bits. This loss for four quantization bits can be practically neglected.
The optimal quantization bit number for a multiple receive SAR system is four bits. Using less quantization bits leads to a SNR_o loss for the resulting compressed SAR image. The low bit quantization noise is applicable for cases where the receive signal contains high amount of additive white noise. The SNR_i of the receive signal should be lower than 0 dB. In this case the loss of 2 dB for single bit quantization must be accounted for. This is not acceptable for a satellite SAR system.

6 Sliding Spotlight Mode

In previous chapters the aspects of hardware design of a synthetic aperture radar had been described. The influence of the receiver front-end, the antenna array position error and quantization had been investigated. In this chapter to processing of the radar data is described.
A SAR system can operate in various operating modes which determine the acquisition geometry. Through different steering of the transmit and the receive antenna beam the resolution of the resulting radar image or the coverage of the illuminated swath can be optimized to match specific requirements. The strip-map mode and the spotlight mode are two most known operating modes for the synthetic aperture radar (SAR) [12]. In the strip-map mode the antenna is fixed in one position with constant incidence angle and squint angle during the whole data acquisition. The radar image in the strip-map mode has an azimuth resolution equal to half the effective antenna length. The ground swath in azimuth direction is illuminated continuously so the azimuth length is not limited by the acquisition geometry. The spotlight mode is optimized regarding azimuth resolution. The radar antenna beam is steered to achieve continuous illumination of a single spot on the ground over the whole data acquisition time. This extends the illumination time of targets in this ground spot. Through this approach the synthetic aperture is effectively much longer. This improves significantly the azimuth resolution of the resulting radar image. The disadvantage of the spotlight mode is the limited length of the illuminated swath in azimuth direction.
The sliding spotlight mode, also called hybrid strip-map/spotlight mode, is a high-resolution operating mode of recently launched spaceborne SAR systems, e.g. TerraSAR-X launched in 2007. This mode is a trade-off between the high resolution of the spotlight mode and the long illumination swath of the strip-map mode [4], [53]. Like in the spotlight mode the radar antenna is steered during the illumination time around a point called steering point. This point is positioned on the illuminated area for the spotlight mode. In the sliding spotlight mode this point is positioned farther from the radar than the illuminated ground area.

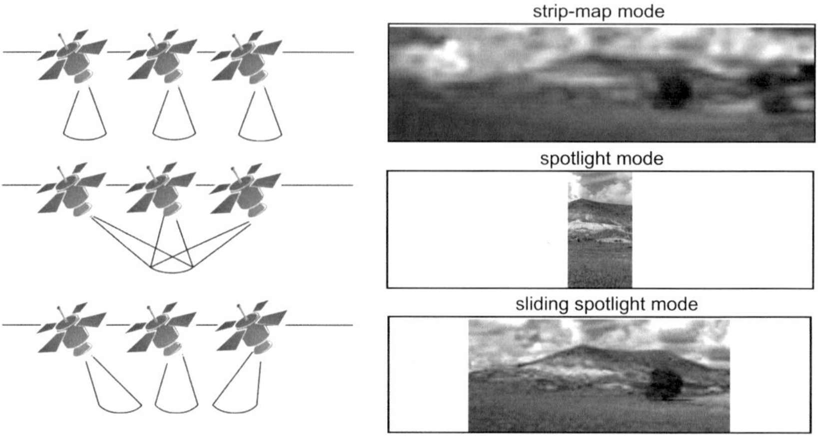

Figure 6.1: Illustration of differences in azimuth resolution and azimuth ground coverage between strip-map mode, spotlight mode and sliding spotlight mode with an optical image.

An illustration of the dependence between azimuth resolution and azimuth coverage is shown in Fig. 6.1. A schematic view of the antenna beam steering for the strip-map mode, the spotlight mode and the sliding spotlight mode is presented in the left side of Fig. 6.1. On the right side an optical image which illustrates the properties regarding image resolution and ground coverage of the given mode is presented. The strip-map mode, where the radar antenna is not steered during data acquisition, is the first from the top. The azimuth resolution in this mode is limited to half the effective antenna length. The swath width (azimuth) is not limited by the acquisition geometry. This allows illumination of very long swaths in azimuth. The azimuth resolution of a SAR image can be improved by steering the antenna beam during the acquisition of the raw data. This method is used in the spotlight mode shown in the middle row in Fig. 6.1. The radar antenna is steered to point to a single ground spot over the whole illumination time. The total illumination time of ground targets is increased with the help of this steering. The azimuth resolution improves significantly which can be observed in Fig. 6.1 as a better resolved image. Through the beam steering the illuminated area in the flight direction (azimuth) is limited. In the optical image only a small part of the picture is visible. The sliding spotlight mode presented at the bottom is a compromise between the wide swath of the strip-map mode and the high resolution of the spotlight mode. This is illustrated by the optical image on the left side of the Fig. 6.1. The resolution is better than in the strip-map mode but not as good as in the spotlight mode. On the other hand the extension of the illuminated scene is larger than in the spotlight mode but not as wide as in the strip-map mode.

6.1 Sliding Spotlight Mode Acquisition Geometry

The sliding spotlight mode, as mentioned before, is a compromise between the high resolution of the spotlight mode and the long illumination swath of the strip-map mode. Similar to the spotlight mode the radar antenna in the sliding spotlight mode is steered during the illumination time. Fig. 6.2(a) shows a three dimensional view of the acquisition geometry of the sliding spotlight mode. Fig. 6.2(b) shows a cut in slant range of the sliding spotlight acquisition geometry. X_I is the flight trajectory during the raw data acquisition, X is the length of the antenna footprint on the ground and X_g is the length of the illuminated ground swath during the data acquisition. The antenna beam for all acquisition times of the radar system is pointing at the same steering point marked on Fig. 6.2(b) with a dot. The steering of the radar beam in the sliding spotlight mode is described in this work by a parameter called A, [94]. The parameter A can be defined as the ratio between the radar platform velocity v_p and the velocity of the footprint of the antenna beam on the ground v_f, Eq. (6.1). Another possibility is to define this parameter as the ratio of the shortest distance between the steering point and the illuminated ground r_1 to the distance between the steering point and the radar antenna $(r_0 + r_1)$, see Fig. 6.2 (b). The r_0 is the smallest distance between the sensor and the illuminated ground. Both definitions are equivalent for a flat Earth geometry, which is a good approximation for an airborne SAR system.

$$A = \frac{v_f}{v_p} = \frac{r_1}{r_1 + r_0} \tag{6.1}$$

The strip-map mode and the spotlight mode can be seen as two limiting cases of the sliding spotlight mode. They can be also easily described with the

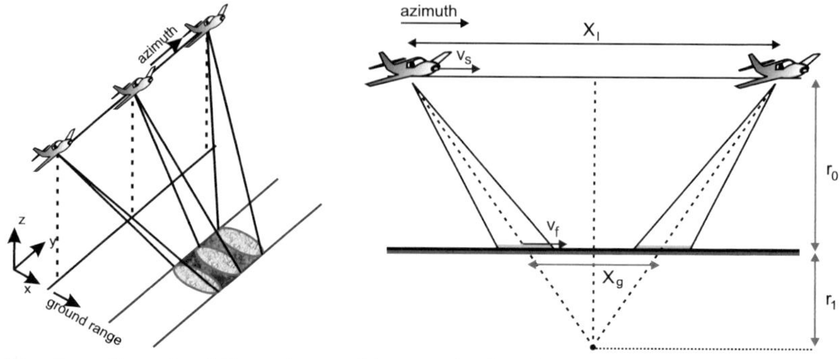

(a) a three dimensional illustration

(b) a cut in the slant range direction through the middle of the swath

Figure 6.2: Raw data acquisition geometry of sliding spotlight mode.

A	SAR mode
0	spotlight mode
$0 < A \leq X/X_I$	semi sliding spotlight mode
$X/X_I < A < 1$	actual sliding spotlight mode
1	strip-map mode

Table 6.1: Parameter A for various SAR modes.

parameter A. In the strip-map mode the velocity of the antenna footprint on the ground v_f is equal to the platform velocity v_p and A is equal to one. In the spotlight mode the velocity of the footprint is zero and the parameter A is also zero.

In dependence on the value of A four different cases of the sliding spotlight mode can be distinguished. The first two were already mentioned. If A takes value 1 or 0 the SAR system works in the strip-map mode or in the spotlight mode adequately. The third case is when A takes values from the interval $(0, \frac{X}{X_I})$. In this case the point in the middle of the swath is resolved with a bandwidth equal to the one of the spotlight mode. Points near edges of the swath in azimuth direction are resolved with smaller bandwidth. This results of various illumination times for targets positioned in the center of the swath and the ones positioned near edges. The azimuth resolution varies with the azimuth position of the illuminated target. The border value is A equal to $\frac{X}{X_I}$. Then only the middle points of the swath in azimuth (one for each range) have the same bandwidth as in the spotlight mode. The last case is when A takes values from the interval $(\frac{X}{X_I}, 1)$. Then radar system works with the actual sliding spotlight mode and all points in the illuminated area are resolved with the same bandwidth. The azimuth resolution is constant over the illuminated swath length. In the following only this case is considered. The azimuth bandwidth for a single point target in the sliding spotlight mode is equal to:

$$B_{az} = \frac{2v_p}{lA} \tag{6.2}$$

Where l is the length of the radar antenna. The azimuth bandwidth increases by the factor A compared to the azimuth bandwidth for a system with the same parameters but working in the strip-map mode. The azimuth resolution is then improved by the same factor A in comparison to the strip-map mode.

$$\delta_{az} = \frac{l}{2}A \tag{6.3}$$

The improvement in the azimuth resolution between the strip-map and the sliding spotlight mode is illustrated with two radar images in Fig. 6.3. The scene was simulated twice with the same parameters of the SAR system. First the raw data was generated and compressed for the strip-map mode. The resulting image for this simulation is presented in Fig. 6.3(a). The distance between point targets is about 1.2 m in both directions. The bandwidth of the transmitted signal is 300MHz, which gives 0.5m resolution in range direction. The antenna length l was chosen 3.33 m. This gives an azimuth resolution of about 1.66m for the strip-map case. This is not enough to resolve point targets in azimuth in the illuminated scene. This is illustrated in Fig. 6.3(a) where points are melting together in azimuth direction so that separate point targets are not distinguishable. The scene was simulated a second time for the sliding spotlight mode with $A = 0.3$. The azimuth resolution in this case is about 0.5 m. The resulting image is presented in Fig. 6.3(b). All point targets in this image are clearly visible. The remaining 'smearing' occurring in the image can be significantly reduced by adequate windowing.

The overall signal bandwidth in azimuth for the whole illuminated swath for

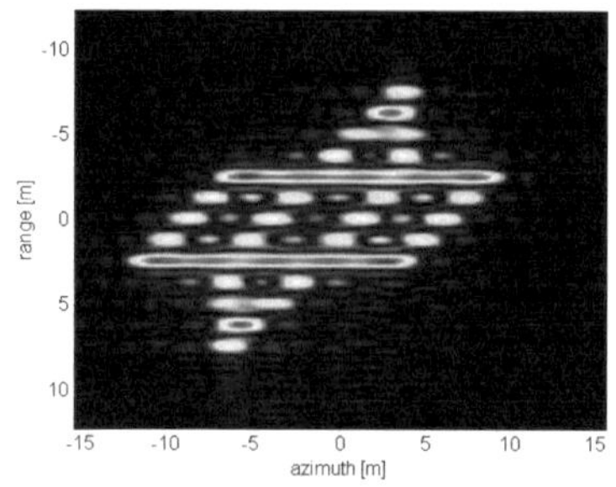

(a) strip-map mode

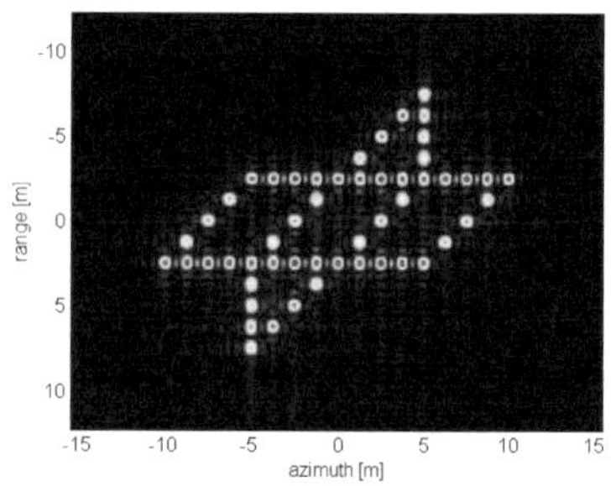

(b) sliding spotlight mode

Figure 6.3: Radar image of a scene illuminated with the same SAR system in strip-map and sliding spotlight mode.

the sliding spotlight mode case can be calculated according to Eq. (6.4)

$$B_{overall} = \frac{2v_p}{l} + \frac{2v_p}{l}\frac{X_I}{X}(1-A) \tag{6.4}$$

In general the azimuth bandwidth of the acquired raw data depends on the PRF (Pulse Repetition Frequency) and the velocity of the radar platform. B_{sa} is the bandwidth of the sampled radar data according to the Nyquist theorem. The sampling in azimuth direction to obtain an unambiguous radar image must fulfill the Nyquist theorem. The requirement for the strip-map mode is described by Eq. (6.5).

$$B_{sa} = \frac{1}{PRF} \geq \frac{2v_p}{l} \tag{6.5}$$

The azimuth sampling in the sliding spotlight mode is identical as for the strip-map mode. It is to be noted that in this case undersampling occurs as the overall bandwidth $B_{overall}$ is higher than B_{sa}. In order to sample the data without any undersampling the sampling frequency should be increased at least by the factor expressed in Eq. (6.6).

$$\frac{B_{overall}}{2/l} = 1 + \frac{X_I}{X}(1-A) \tag{6.6}$$

The undersampling in the sliding spotlight mode caused by choosing a lower PRF can be resolved by an adequate signal processing. The processing and necessary conditions are described in following sections of this chapter. A limit value of the parameter A to which the signal is not undersampled A_L can be calculated as in Eq. (6.7), [49].

$$A_L = 1 + \frac{X}{X_I} \frac{B_{sa}/v_p - 2/l}{2/l} \tag{6.7}$$

6.2 Sliding Spotlight Mode Raw Data

The radar system sends chirp signals at times $t = -\frac{t_n}{2}$, where t_n is the half of the transmitted pulse duration τ. The transmitted chirp signal is described by Eq. (6.8)

$$s_{Tx}(t) = \exp\left(-j\pi k_r \left(t - t_n\right)^2\right) \cdot \mathrm{rect}\left\{\frac{t - t_n}{\tau}\right\} \tag{6.8}$$

Where c_0 is the light velocity and k_r is the chirp rate, which is defined according to Eq. (6.9).

$$k_r = \frac{B}{\tau} \tag{6.9}$$

B is the bandwidth of the transmitted signal. The radar signal is reflected back from a point target positioned at the illuminated swath. The minimal distance between the illuminated point target and the radar antenna is r_0. The point target can be 'seen' by the radar antenna as long as the condition in Eq. (6.10) is fulfilled.

$$|Au - u_{target}| \leq \frac{X}{2} \tag{6.10}$$

u is the azimuth direction corresponding to the flight trajectory which can be easily recalculated into slow time. For a scene geometry as presented in Fig. 6.2 u corresponds to y coordinate. u_{target} is the position of the point target in azimuth direction. The position of the point target can be described by coordinates (r_0, u_{target}) in slant range–azimuth coordinates. The receive signal is described by Eq. (6.11).

$$\begin{aligned} s_{Rx}(t,u) &= \exp\left(-j\omega\frac{2r}{c_0}\right)\cdot\exp\left(-j\pi k_r\left(t-t_n-\frac{2r}{c_0}\right)^2\right)\cdots \\ &\quad \mathrm{rect}\left\{\frac{t-t_n-\frac{2r}{c_0}}{\tau}\right\}\cdot\mathrm{rect}\left\{\frac{Au-u_{target}}{X}\right\} \end{aligned} \tag{6.11}$$

The changing distance r between the point target and the radar antenna phase center can be calculated according to Eq. (6.12).

$$r=\sqrt{r_0^2+u^2}=\sqrt{r_0^2+(v_pt)^2} \tag{6.12}$$

A more complex scene can be seen as a superposition of many point targets.

6.3 Spectrum of the Sliding Spotlight Mode Raw Data

In the strip-map mode the radar signal is sampled in the azimuth direction with a certain PRF. The minimal value of PRF_{min} can be estimated from the geometry of the radar system and the velocity of the radar platform v_p. The value can be calculated according to Eq. (6.13). This minimal frequency is sufficient to sample the radar signal in azimuth direction without ambiguities. In practical applications a certain oversampling is applied, usually around 1.2. The radar antenna in the strip-map mode is not steered during the data acquisition so every target at the same slant range has the same Doppler history.

$$\mathrm{PRF}_{min}=\frac{2v_p}{l} \tag{6.13}$$

For the spotlight mode the radar antenna is steered towards a single point on the ground. This results in a longer Doppler history for every target in the illuminated spot. Similar in the sliding spotlight mode the radar antenna is steered after a steering point but positioned further away. The radar antenna in the sliding spotlight mode is steered 'slower' than in the spotlight mode, see Fig. 6.2. The Doppler history of a point target depends on its position in azimuth direction u_{target}. Fig. 6.4 shows the change of the Doppler frequency in dependence on slow time (azimuth direction).

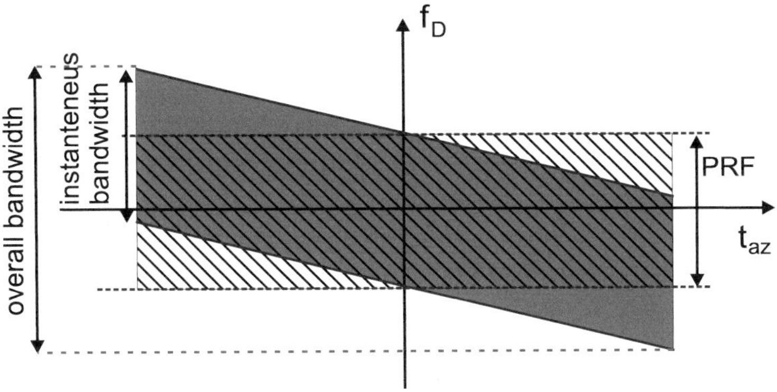

Figure 6.4: The Doppler frequency n the sliding spotlight mode in slow time.

The PRF for the sliding spotlight mode is chosen the same as for the strip-map mode. This, as can be seen in Fig. 6.4, results in undersampling and aliasing. The overall bandwidth can be calculated according to Eq. (6.14). The overall bandwidth is higher than the one resulting from PRF. It is to be noted that the instantaneous bandwidth in the sliding spotlight mode is not undersampled.

$$\frac{B_{overall}}{2/l} = 1 + \frac{X_I}{X}(1 - A) \tag{6.14}$$

The azimuth spectrum for a target is centered around the frequency f_{center}, which can be calculated according to Eq. (6.15). This center frequency depends mostly on the position of the illuminated target in azimuth direction u_{target}.

$$f_{center} = 2u_{target}v_p(A - 1)/(\lambda r A) \tag{6.15}$$

The azimuth bandwidth of a single point target in the sliding spotlight mode is equal to:

$$B_{target} = \frac{2v_p}{lA} \tag{6.16}$$

6.4 Processing Algorithms

The two chosen SAR processing algorithms: the chirp scaling algorithm and the omega-K algorithm are described in this section. In the literature also other solution can be found [50], [54], [70], [71], [104]. The specific feature of the sliding spotlight mode is the undersampling of the raw data in azimuth direction. This might result in aliasing in azimuth direction in the compressed radar image. This issue is resolved with the help of an azimuth filter. Further the radar data can be processed with a modified strip-map algorithm. The chirp scaling algorithm and the omega-K algorithm are modified accordingly to the enable processing of the sliding spotlight mode data.

6.4.1 Azimuth Filter

The algorithm for processing SAR data in the sliding spotlight mode must overcome the problem of the undersampled raw data and, caused by this, the aliasing in azimuth direction. A possible solution to this problem is an azimuth filter applied on the raw data in a pre-processing step [22], [49]. In the sliding spotlight mode the PRF value is chosen the same as it would be in case of the strip-map mode. An aliasing effect in azimuth direction occurs because the overall bandwidth of the sliding spotlight mode is larger than the used PRF. The azimuth filter with three steps resamples the radar data and by this resolves the problem of undersampling. This is possible, as the instantaneous bandwidth is not higher than the one resulting from azimuth sampling. The angle under which the scatterers are illuminated changes over the acquisition time so that the Doppler history is shifted in frequency adequately to the position of the scatterer.
The azimuth filter resolving the aliasing of the raw data is defined according to Eq. (6.18). The raw data is partially compressed and then resampled. Resulting is a partially compressed radar data with new length in azimuth direction and new sample spacing.

$$\tilde{f}(n\Delta u_2, u', r) = \exp\left(j2\pi\frac{u'^2}{\lambda\tilde{r}}\right)\cdot \ldots \qquad (6.17)$$
$$FFT\left\{f(i\Delta u - u, u, r)\cdot\exp\left(j2\pi\frac{u^2}{\lambda\tilde{r}}\right)\right\}$$

For $i = -U/2 \ldots U/2-1$ and $n = -P/2 \ldots P/2-1$, λ is the wavelength and $\tilde{r}$ is the reference range for the azimuth filter. Δu is the original sample spacing in azimuth direction. The azimuth vector u is of the length U. Δu_2 is the new sampling spacing in azimuth direction of the azimuth vector u' with the length of P. Fig. 6.5 shows the schematic of the azimuth filter. The

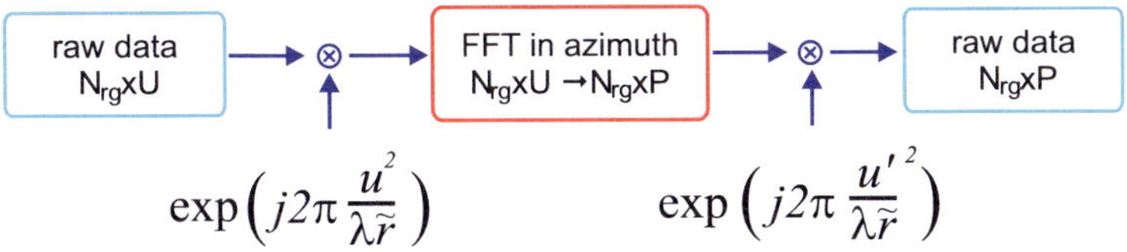

Figure 6.5: Block diagram of the azimuth filter.

received radar raw signal is first multiplied with an exponential function, as shown in Fig. 6.5. This reduces the slope of the azimuth chirps in the raw data. A Fourier transformation in azimuth direction is applied to the signal with a new length of P. The resampling of the radar data is implemented as zero-padding and following Fourier Transformation. The radar signal is then multiplied with a second exponential function with a new sampling spacing Δu_2 of the azimuth vector u'.

The function of the azimuth filter is illustrated in Fig. 6.6. The scenario considered here consists of three point targets each marked with a red cross. Point targets are positioned at the beginning, in the middle and at the end of the illuminated swath in azimuth direction. The consecutive time-Doppler graphs in Fig. 6.6 show three steps of the azimuth filter. The first graph in this figure shows the Doppler history for all three targets before the azimuth filter is applied. The figure presents the momentary Doppler frequency f_D of the receive signal for all three targets plotted over slow time (azimuth time). It is to be noted that the Doppler history varies for each point target. The Doppler history depends on the azimuth position of the target for a sliding spotlight mode. The unambiguous spectrum is indicated with a yellow rectangle. The second graph in Fig. 6.6 shows the momentary Doppler frequency after the first multiplication. The slope of the Doppler history in slow time is reduced to fit in the unambiguously sampled area. The third graph illustrates the resampling of the raw data. The new azimuth bandwidth ensures unambiguous sampling of the data.

The new sampling rate of the raw data in azimuth direction is to be chosen according to Eq. (6.18), [49].

$$P \geq \frac{\lambda \tilde{r}}{2\Delta u} B_{overall} \tag{6.18}$$

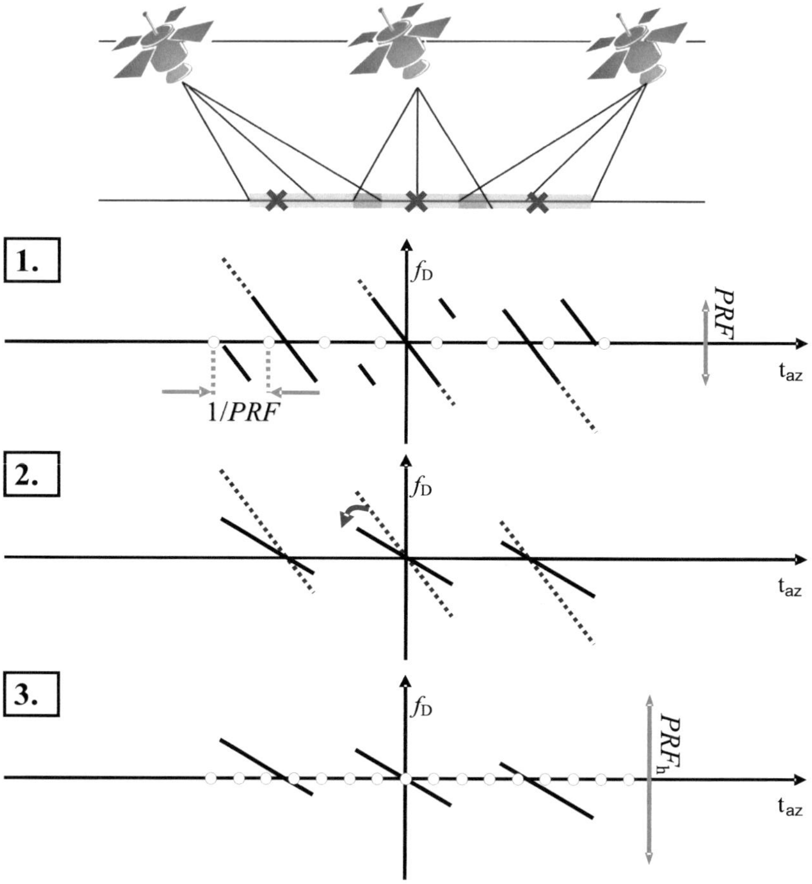

Figure 6.6: Schematic illustration of azimuth filter.

The new sample spacing must fulfill the Nyquist criterion for the overall bandwidth:

$$\Delta u_2 \leq \frac{1}{B_{overall}} \tag{6.19}$$

$\tilde{r}$, which is the reference range for the azimuth filter, can be chosen freely from the values fulfilling Eq. (6.20). The chosen value of $\tilde{r}$ has no significant influence on the outcome of the algorithm. In following $\tilde{r}$ is calculated for r equal to r_s for the middle of the illuminated swath.

$$\tilde{r} = \frac{r}{1-A} \quad \text{where} \quad r \in (r_{min}, r_{max}) \tag{6.20}$$

r_{min} and r_{max} specify the minimal and maximal range of the illuminated swath. After applying the azimuth filter the signal has a new azimuth sample spacing and a new number of samples, which allows the processing of the sliding spotlight mode raw data with algorithms for the strip-map mode. A change in the compression algorithm is still necessary in order to compensate for the azimuth filtering. That will be described in further section describing signal processing algorithms. The maximal azimuth frequency of the radar signal after the azimuth filter be calculated with Eq. (6.21).

$$\max(f_{az}) = \frac{v_p}{2\Delta u_2} \tag{6.21}$$

The parameters of an equivalent strip-map mode SAR system l_{hyb} and X_{hyb}, which would have the same azimuth resolution as a SAR system working in the sliding spotlight mode, are expressed in Eq. (6.22) and Eq. (6.23).

$$l_{hyb} = \frac{2}{B_{az}} \tag{6.22}$$

$$X_{hyb} = \frac{\lambda r_s}{l_{hyb}} \tag{6.23}$$

6.4.2 Chirp Scaling Algorithm for the SAR Sliding Spotlight Mode

The chirp scaling algorithm is a processing algorithm for the SAR raw data [76]. The chirp scaling algorithm consists of Fourier transformations and multiplications with exponential functions. No interpolation of the radar data is necessary. This makes the algorithm fast and easy to implement. In the following the chirp scaling algorithm is described. Fig. 6.7(a) shows a block diagram of the chirp scaling algorithm for the sliding spotlight mode proposed in this work [65]. Additional blocks necessary for the sliding spotlight mode are marked in red. The additional blocks are the azimuth filter at the beginning of the algorithm and the correction for the azimuth filter in further processing. The additional blocks are introduced to the algorithm when undersampling of the radar raw data occurs. The threshold for no undersampling of the raw data is $A = A_L$. It is to be noted that after injecting the azimuth filter into the data processing chain some parameter values must be updated. The azimuth filter described in the previous section is applied to the raw data before the actual compression algorithm, see Fig. 6.7(a). Next the radar raw data is transformed from the time domain to the range-Doppler domain through Fourier transformation in the azimuth direction. The radar data is multiplied with the first exponential function $\Phi_1(t_{rg}, f_{az}, r_{ref})$. This multiplication applies the chirp scaling to the radar data. The range curvatures for all targets regardless their range position are the same after this multiplication. The first exponential function is described by Eq. (6.24).

$$\begin{aligned} \Phi_1(t_{rg}, f_{az}, r_{ref}) &= \exp(-j\pi \cdot K_s(f_{az}, r_{ref}) \cdot \ldots \\ & \quad C_s(f_{az}) \cdot [t_{rg} - t_{ref}]^2) \end{aligned} \tag{6.24}$$

Where t_{rg} is the range time (fast time). The range time and slant range can be easily recalculated according to Eq. (6.25).

$$t_{rg} = \frac{2}{c_0} \cdot r \tag{6.25}$$

t_{ref} is the reference time defined as:

$$t_{ref}(f_{az}) = \frac{2}{c_0} r_{ref}[1 + C_s(f_{az})] \tag{6.26}$$

As the reference range r_{ref} usually chosen in the center of the illuminated swath. $C_s(f_{az})$ is the curvature factor. It can be calculated according to Eq. (6.27).

$$C_s(f_{az}) = \frac{1}{\sqrt{1 - \left(\frac{\lambda f_{az}}{2v_p}\right)^2}} - 1 \tag{6.27}$$

$K_s(f_{az}, r_{ref})$ is the effective FM chirp rate in range and can be calculated according to Eq. (6.28).

$$K_s(f_{az}, r_{ref}) = k_r \left(1 + k_r r \frac{2\lambda}{c_0^2} \frac{\left(\frac{\lambda f_{az}}{2v_p}\right)^2}{\left[1 - \left(\frac{\lambda f_{az}}{2v_p}\right)^2\right]^{3/2}} \right)^{-1} \tag{6.28}$$

Where f_{az} is the azimuth frequency and f_{rg} is the range frequency.
Next step in the chirp scaling algorithm is a Fourier transformation in range direction. Now the radar data is transformed into the two dimensional frequency domain. Second multiplication with an exponential function $\Phi_2(f_{rg}, f_{az}, r_{ref})$ follows. The second multiplication applies the bulk RCMC and range focus into the radar data. The function Φ_2 can be calculated according to Eq. (6.29).

$$\begin{aligned} \Phi_2(f_{rg}, f_{az}, r_{ref}) &= \exp\left(-j\pi \frac{f_{rg}^2}{K_s(f_{az}, r_{ref}) \cdot (1 + C_s(f_{az}))} \right) \cdot \ldots \\ & \exp\left(j\frac{4\pi}{c_0} f_{rg} \cdot r_{ref} \cdot C_s(f_{az}) \right) \end{aligned} \tag{6.29}$$

As the next step a correction for the azimuth filter is applied to the radar data. The azimuth filter already partially compressed the raw data. This must be accounted for in the processing algorithm. The multiplication with the exponential correction function is necessary for accurate compression of the radar image. The correction function is described with Eq. (6.30).

$$\Phi_{corr}(f_{az}) = \exp\left(j2\pi \left(\frac{f_{az}}{v_p}\right)^2 \frac{\lambda \tilde{r}}{4} \right) \tag{6.30}$$

After the multiplication with $\Phi_{corr}(f_{az})$ the radar data is transformed back into the range-Doppler domain with an inverse Fourier transformation in range direction. The last multiplication with the third exponential function $\Phi_3(t_{rg}, f_{az})$ introduces the azimuth focus and phase correction. The third function is described by Eq. (6.31).

$$\begin{aligned} \Phi_3(t_{rg}, f_{az}) &= \exp\left(-j\frac{2\pi}{\lambda} c_0 \tau \left(1 - \left(1 - \left(\frac{\lambda f_{az}}{2v_p} \right)^2 \right)^{1/2} \right) \right) \cdot \ldots \\ & \exp\left(j\theta_\Delta(f_{az}, r) \right) \end{aligned} \tag{6.31}$$

Where $\theta_\Delta(f_{az}, r)$ is calculated according to Eq. (6.32).

$$\theta_\Delta(f_{az}, r) = \frac{4\pi}{c_0^2} K_s(f_{az}, r_{ref}) \cdot (1 + C_s(f_{az})) \cdot C_s(f_{az}) \cdot (r - r_{ref}) \quad (6.32)$$

The last step is the inverse Fourier transformation in azimuth. A fully compressed radar image is provided after this transformation.

6.4.3 Omega-K Algorithm for the SAR Sliding Spotlight Mode

The second SAR processing algorithm considered in this work is the omega-K algorithm [12]. The name comes from the fact that the radar data is processed in the two dimensional frequency domain. The range frequency is often described with the Greek letter ω and the azimuth frequency with the wavenumber identified with the letter K. In this work f_{rg} and f_{az} is used to describe the range frequency and the azimuth frequency accordingly K and ω. These frequencies can be derived from range time (fast time) t_{rg} and azimuth time (slow time) t_{az}. This notation is the same as the one used in the description of the chirp scaling algorithm in previous section.
The block diagram of the omega-K algorithm for the sliding spotlight mode as proposed in this work [105] is presented in Fig. 6.7(b). Modifications, necessary to process the sliding spotlight mode are marked in red. The aliasing in azimuth direction in the received data occurs when A is smaller than A_L. In this case the azimuth filter is applied on the radar raw data. It is to be noted that some of the parameters used in following processing are recalculated to match new system parameters. The SAR raw data is transformed into the two dimensional frequency domain with a two dimensional Fourier transform, see Fig. 6.7(b). The next step is a multiplication with the two dimensional reference function, which is described by Eq. (6.33). The reference function applies the bulk azimuth compression, range cell migration correction (RCMC) and rage compression on the radar data. The azimuth compression is calculated for the reference range r_{ref}. This means the points at the reference range will be fully focused. Other points not positioned at the reference range will be slightly defocused. The defocusing will increase with increasing distance from the reference range.

$$\Phi_{ref}(f_{rg}, f_{az}) = \exp(j\theta(f_{rg}, f_{az})) \quad (6.33)$$

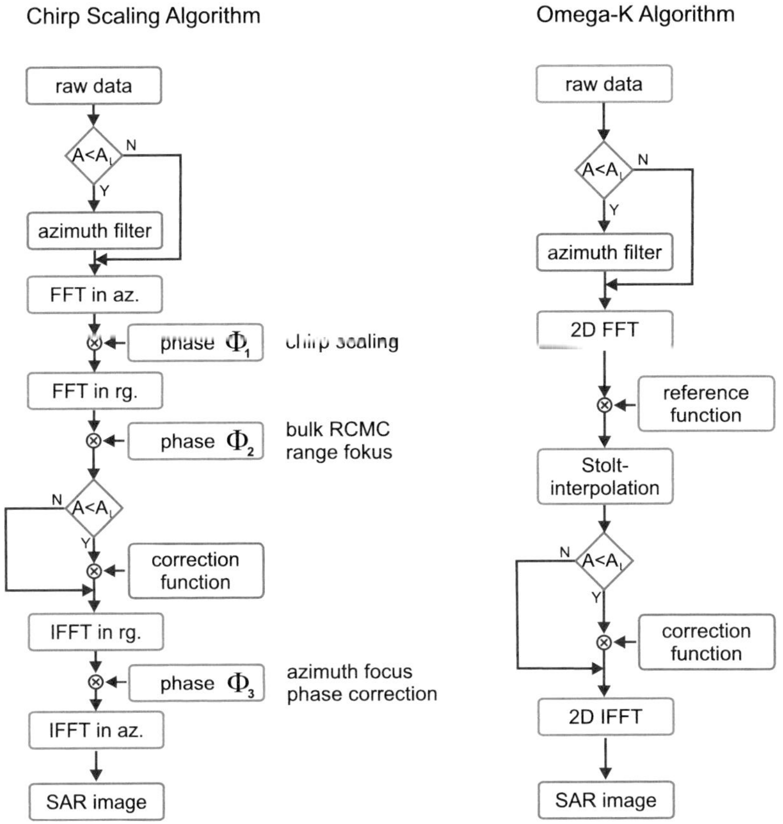

(a) chirp scaling algorithm (b) omega-K algorithm

Figure 6.7: Block diagrams of SAR processors for sliding spotlight mode.

$$\theta(f_{rg}, f_{az}) = \frac{4\pi r_{ref}}{c_0}\sqrt{(f_{rg} + f_c)^2 - \frac{c_0^2 f_{az}^2}{4v_p^2}} + \frac{\pi f_{rg}^2}{k_r} \tag{6.34}$$

The following step of the omega-K algorithm fully focuses the data also for other ranges. A Stolt interpolation is applied to the radar data. The interpolation equation can be seen in Eq. (6.35). The variable for the range frequency f_{rg} is substituted by the new variable f'_{rg}.

$$\sqrt{(f_{rg} + f_c)^2 - \frac{c_0^2 f_{az}^2}{4v_p^2}} = f_c + f'_{rg} \tag{6.35}$$

f_c is the center frequency, where $f_c = \frac{\omega_c}{2\pi}$. Next the radar data is multiplied with the exponential correction function. This multiplication removes the influence of the azimuth filter. The exponential function is calculated according to Eq. (6.30). Last step of the omega-K algorithm is the transformation back into the two dimensional time domain through two dimensional Fourier transformation. A fully compressed radar image is provided after this step.

6.5 Antenna Pattern Compensation

The antenna pattern of the transmit and the receive antenna have not been considered in previous sections describing the signal processing. The pattern was modeled as a rectangular window with the length of the antenna footprint on the ground X. This model is very simple and sufficient for developing and verifying general SAR processing algorithms. In the processing of real radar data a more adequate model of the antenna pattern is necessary. A more realistic characteristic of the transmit and the receive antenna will be assumed and introduced into the data processing chain in this section. Also a compensation of the antenna pattern for the strip-map and the spotlight mode is described. A compensation algorithm for the antenna pattern for the sliding spotlight mode is proposed. A similar procedure can be applied for the windowing of the radar data to reduce side lobes level.

The radar transmits a signal with a transmit antenna and receives the reflected echo with a receive antenna. The radar receive signal has the transmit antenna characteristic w_{Tx} and the receive antenna characteristic w_{Rx}. Both characteristics can be easily recalculated for the SAR system to be functions dependent on the fast-time and azimuth position of the radar antenna. The angle of the incoming reflection can be calculated assuming flat Earth model. The receive signal for the strip-map mode including antenna patterns can be calculated according to Eq. (6.36).

$$\begin{aligned} s'_{Rx}(t_{rg},u) &= w_{Tx}\left(t_{rg}-\frac{2r}{c_0},u\right)\cdot s_{Rx}\left(t_{rg}-\frac{2r}{c_0},u\right)\cdots \\ & \quad w_{Rx}\left(t_{rg}-\frac{2r}{c_0},u\right) \end{aligned} \tag{6.36}$$

The characteristic of the antenna can be compensated in the SAR processing. To simplify the notation it will be assumed further that both antennas for transmit and for receive have the same characteristic.

$$w_{Tx/Rx}(t_{rg},u) = w_{Tx}(t_{rg},u) = w_{Rx}(t_{rg},u) \tag{6.37}$$

The received signal in the strip map mode can be then described as:

$$s'_{Rx}(t_{rg},u) = w^2_{TxRx}\left(t_{rg}-\frac{2r}{c_0},u\right)\cdot s_{Rx}\left(t_{rg}-\frac{2r}{c_0},u\right) \tag{6.38}$$

The antenna pattern compensation for the strip-map, spotlight and sliding spotlight mode is described in following.

6.5.1 Antenna Pattern Compensation for the Strip-Map Mode

All targets in the illuminated swath for a specified range and height have a similar Doppler characteristic in the strip-map mode. Targets are 'seen' in azimuth direction under the same angles by the radar antenna. The weighting resulting from the antenna pattern is the same for all targets independent on their azimuth position. Eq. (6.38) which corresponds to the strip-map mode shows that the antenna pattern weighting is independent of the position of the target in azimuth direction. The radar data is a superposition of many reflections of various targets weighted with the transmit and receive antenna pattern. Received echoes from targets positioned in various azimuth positions are recorded in corresponding azimuth cells in the raw data. The spectrum of all targets responses lies in the same boundaries. This means that every sample in the compressed data spectrum is a superposition of all targets in the illuminated swath weighted with the same part of the antenna pattern.

Compensation of the antenna pattern in azimuth in the frequency domain is quite simple to implement.
For the following examples a sinusoidal antenna radiation pattern is assumed. Eq. (6.39) shows a one dimensional two-way antenna pattern in azimuth direction.

$$w^2_{TxRx}(u) = \text{sinc}^2\left(\frac{\pi X \theta(u)}{\lambda}\right) \tag{6.39}$$

$\theta(u)$ is defined according to Eq. (6.40).

$$\theta(u) = \text{asin}(v_p u / r) \tag{6.40}$$

Fig. 6.8 shows the spectrum of a compressed image of a single point target. The antenna radiation pattern in this example is cut to the synthetic aperture length to better illustrate the compensation. The image on the left side in Fig. 6.8 shows a two dimensional spectrum of a compressed point target. A cut through the azimuth frequency is shown on the right side of the figure. The cut is through the middle of the spectrum. The point target is weighted with the azimuth two way antenna pattern. The influence on the spectrum of the target can be clearly seen in both pictures.

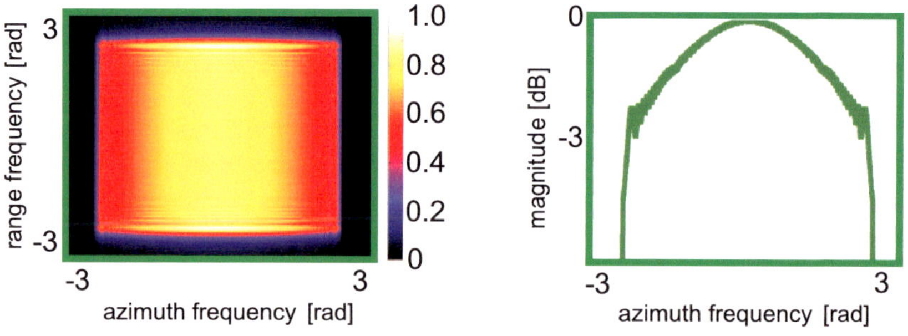

Figure 6.8: The influence of the antenna pattern on a spectrum of a compressed point target in the strip-map mode.

The antenna pattern, scaled for the radar data extension, can be calculated according to Eq. (6.41). The recalculation scales the antenna pattern accordingly to the extension of the radar signal bandwidth and the sampling

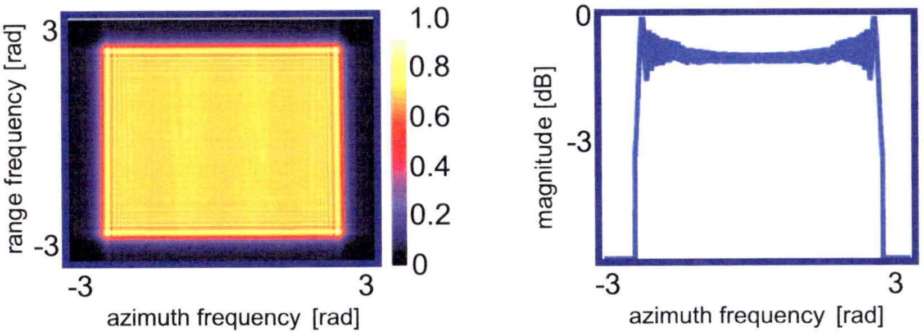

Figure 6.9: Spectrum of a compressed point target in strip-map mode with compensated antenna pattern.

frequency.

$$w_{TxRx}\left(\frac{X\theta(u)}{0.886\lambda}\right) = W_{TxRx}\left(\frac{Xf_{az}}{2v_p}\right) \tag{6.41}$$

The radar data is multiplied with the inverse of the scaled antenna pattern in the frequency domain. The result can be seen in Fig. 6.9. The two dimensional spectrum of the compressed point target is presented on the left side. A cut in the azimuth direction is shown on the right side. The weighting caused by the antenna radiation pattern in azimuth direction has been fully compensated.

The compensation of the antenna pattern in range direction is achieved with a different approach. The radar antenna in elevation is fixed under a specific looking angle. Each range bin is illuminated under a specific angle and by this with a specific fixed weighting from the antenna pattern. This makes a compensation of the antenna pattern in elevation (range) direction easy to implement in the time domain.

The correction in range is independent of the operating mode[1] of the SAR system. In azimuth a different approach must be chosen for each of the strip-map, spotlight and sliding spotlight mode. In the strip-map mode the correction takes place in the frequency domain. The radar data is transformed into frequency domain and is multiplied with an adequate antenna radiation pattern recalculated into the frequency domain. The correction can be applied to

[1] Considered modes are the strip-map, spotlight and sliding spotlight mode.

the radar image or can be applied as an integral part of the radar processing algorithm. It is also possible to choose the antenna pattern in such a way that no compensation in the processing algorithm is necessary. In this case the antenna pattern can take the role of a window. The side lobes for compressed radar image can be reduced by an adequate design of the antenna pattern.

6.5.2 Antenna Pattern Compensation for the Spotlight Mode

The antenna radiation pattern correction for the strip-map mode radar data is usually made in the time domain. Fig. 6.10 shows the geometry of the spotlight mode with two targets marked with green and red colored rectangles. The target marked with green color is positioned in the middle of the illuminated swath in azimuth direction and therefore is illuminated at the maximum of the transmit and receive antenna pattern. The reflected signal from the target in the middle is also received at the middle of the receiver antenna pattern. The target marked with red is positioned at the edge of the illuminated swath.

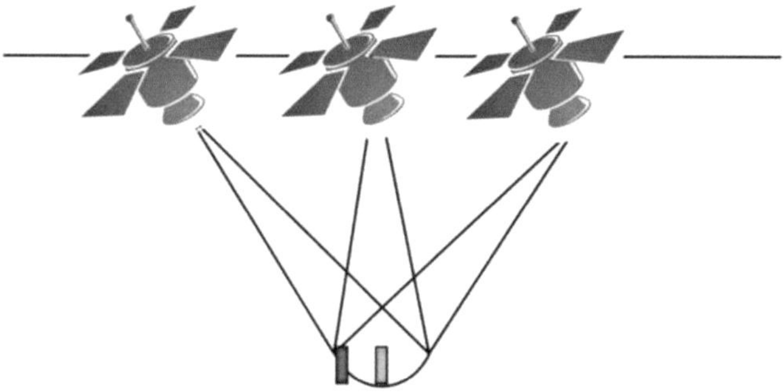

Figure 6.10: The geometry of a SAR system in the spotlight mode with two targets marked one in green in the middle of the illuminated swath and one in red at the edge of the swath.

This means that it is also positioned at the 'edge' of the radiation pattern of the transmitter and receiver antenna. It is to be noted that the weighting caused by the antenna radiation pattern is related to the position in the azimuth direction of the target but it is constant over the illumination time. Therefore the compensation can be applied in the time domain. The radar data in the time domain is multiplied with inverse of the accordingly scaled two-way antenna pattern. This operation can be applied for range and azimuth directions simultaneously.

6.5.3 Antenna Pattern Compensation for the Sliding Spotlight Mode

In the strip-map mode and in the spotlight mode the weighting can be compensated for a radar image in the frequency or in the time domain accordingly. The compensation of the antenna pattern in the sliding spotlight mode is more complex. Fig. 6.11 shows the illumination geometry of the sliding spotlight mode with two point targets on both edges of the swath in azimuth direction marked with red and green rectangles. The two targets are illuminated in different time and under different squint angle resulting in different Doppler spectra. The compensation in the time domain for the sliding spotlight mode would be very complicated as the antenna beam is sliding with time over the ground. The compensation of the antenna pattern in the frequency domain without preprocessing is also complicated as the target's Doppler history depends on its position in azimuth direction. Fig. 6.12 shows the magnitude of the spectrum of a scenario with two point targets positioned as shown in Fig. 6.11. On the left a two dimensional magnitude of the compressed data spectrum in range and azimuth frequency is presented. The spectra from both targets are marked with a green and a red colored frame. A cut in the azimuth frequency is shown on the right in Fig. 6.12. The weighting caused by the antenna radiation pattern can be clearly seen on both pictures. The pattern was set to zero beyond the synthetic aperture limit to make the weighting and compensation more visible.
In this example the position of the two point targets and system parameters are chosen so that the Doppler histories of both targets do not overlap for the purpose of clearer illustration. In the general case the spectrum of a scene consists of multiple overlapping spectra of all illuminated targets.
The compensation of the antenna pattern is possible through a position depen-

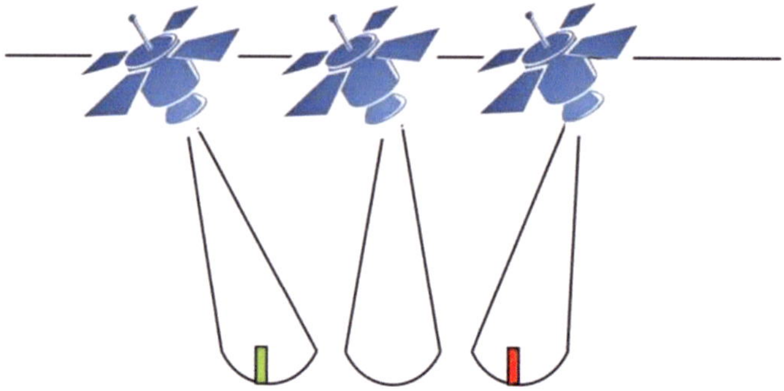

Figure 6.11: The geometry of the sliding spotlight mode with targets marked in green and red on the edges of the swath in azimuth.

dent shift of the target's spectrum. The necessary shift in frequency can be calculated according to Eq. (6.42). The shift depends on the azimuth position of the point target.

$$f_{shift}(t_{az}) = -\frac{v_p^2 t_{az}(A-1)}{\lambda r_{ref} A} \tag{6.42}$$

Fig. 6.13 presents a diagram of the antenna pattern compensation algorithm for the sliding spotlight mode. The data is multiplied with an exponential function in the time domain. This function is described by Eq. (6.43).

$$\begin{aligned} C(t_{az}) &= \exp\left\{j2\pi \cdot f_{shift}(t_{az}) \cdot t_{az}\right\} \\ &= \exp\left\{-j\pi \frac{v_p^2 t_{az}^2 (A-1)}{\lambda r_{ref} A}\right\} \end{aligned} \tag{6.43}$$

Points at different azimuth positions are shifted in frequency by different values, so that the spectrum of each of them is in the same frequency span after the Fourier transformation. The data is then transformed into the frequency domain where the compensation of the antenna radiation pattern is applied. It is necessary to recalculate the antenna pattern to match the extension in

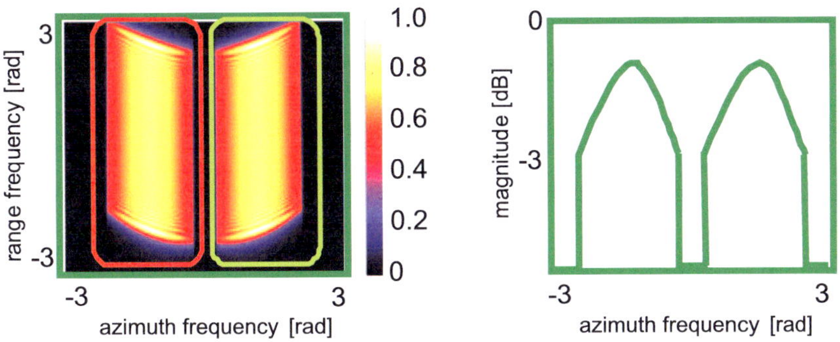

Figure 6.12: Magnitude of the spectrum of two point targets illuminated in the sliding spotlight mode. Two dimensional representation of the magnitude of the spectrum (left) and the cut in azimuth frequency (right).

frequency of the radar data. This can be done through adequate frequency mapping according to Eq. (6.44).

$$W_{TxRx}(f_{az}) = W_{TxRx}\left(\frac{Af_{az}X}{2v_p}\right) \tag{6.44}$$

The capital letter indicates that although the antenna pattern was not directly transformed the compensation is applied in the frequency domain. It is to be noted that the mapping of the antenna pattern depends also on the parameter

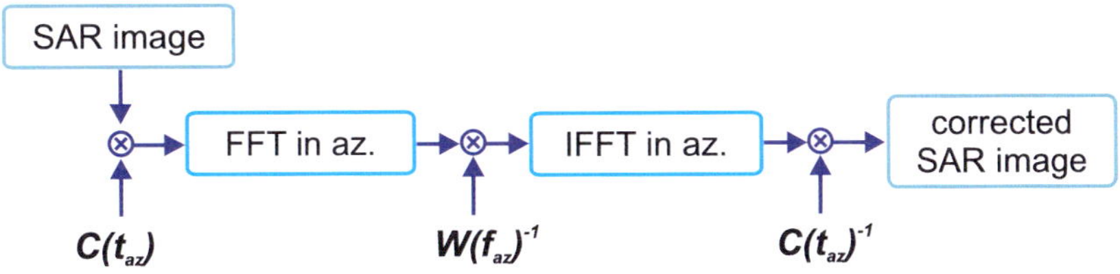

Figure 6.13: Block diagram of the antenna pattern compensation algorithm for sliding spotlight mode.

A. Eq. (6.44) shows the one-way antenna pattern. After compensation the radar data is transformed back into the time domain. An inverse shift in frequency is applied through multiplication with the $C(t_{az})^{-1}$ function, as in Eq. (6.45).

$$C(t_{az})^{-1} = \exp\left\{ j\pi \frac{v_p^2 t_{az}^2 (A-1)}{\lambda r_{ref} A} \right\} \tag{6.45}$$

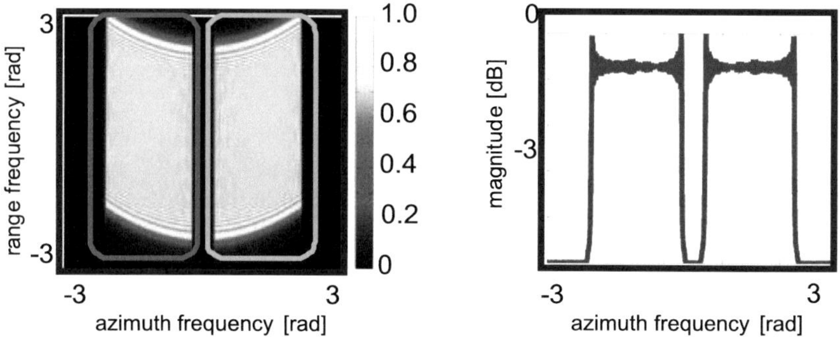

Figure 6.14: Magnitude of the spectrum of two point targets illuminated in the sliding spotlight mode after antenna pattern compensation. Two dimensional representation of the magnitude of the spectrum (left) and the cut in azimuth frequency (right).

The spectrum of two point targets with compensated antenna radiation pattern is presented in Fig. 6.14. The test scene consists of two point targets positioned so that their spectra do not overlap. On the left image a two dimensional representation of the spectrum's magnitude is presented. A cut in the azimuth frequency is shown on the right side. The cut is marked on the left image with an arrow. It can be clearly seen in Fig. 6.14 that after the compensation the magnitude of the spectrum is flat and no antenna pattern influence is present.

It is to be noted that the compensation of the antenna pattern in azimuth for the sliding spotlight mode is applied on the compressed radar image. For both the strip-map mode and the spotlight mode the compensation of antenna pattern can be a integral part of the SAR processing algorithm.

6.5.4 Windowing

It is common for SAR systems to have windowing applied on the focused radar image to reduce the side lobes of the targets response. This improves the quality of the radar image by reducing side lobe levels, which could mask weak targets near a strong scatterer. As a disadvantage a certain widening of the main lobe must be accounted for. The windowing is applied in the frequency domain through multiplication with an adequate windowing function. For the sliding spotlight mode the procedure of windowing in azimuth direction is similar to the compensation of the antenna radiation pattern. The radar image data is multiplied with the exponential function from Eq. (6.43) to shift all targets in the frequency domain to the center of the spectrum. In the next step the radar data is multiplied with the adequately mapped window function in the frequency domain. The data is transformed back into the time domain with the inverse Fourier transformation. Finally targets are shifted back to the original position in the spectrum through multiplication with the exponential function from Eq. (6.45).

6.6 Extension of the Illuminated Swath

The azimuth filter recalculates the radar raw data. The raw data is collected as the radar antenna travels the distance X_I while the swath length X_g is illuminated. The radar data is recalculated by the azimuth filter to a scene of the extension of X_s. This length can be estimated according to Eq. (6.46).

$$X_s = \frac{X_g}{A}\left|1 - (1 - A)\frac{\tilde{r}}{r_s}\right| + \lambda\left|\frac{\tilde{r} - r_s}{lA}\right| \tag{6.46}$$

With $\tilde{r}$ defined as in Eq. (6.20) the Eq. (6.46) simplifies to:

$$X_s = \frac{r}{lA} \tag{6.47}$$

and is independent of the actual value of X_g. If the illuminated scene X_g is wider than X_s an adequate number of zeros must be added at both sides of the data matrix. Otherwise the scene will be wrapped around. Fig. 6.15 shows the principle of adding zeros to the data matrix. The required number of zeros can be calculated with Eq. (6.48).

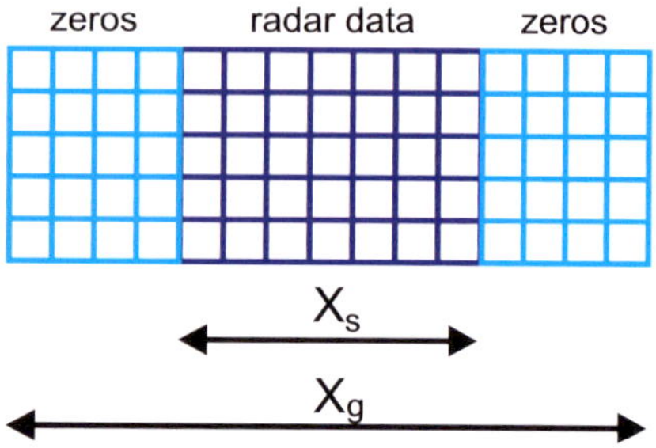

Figure 6.15: Data matrix after azimuth filter and with zeropadding.

$$U_{zero} = \frac{X_g - X_s}{2\Delta u_2} \tag{6.48}$$

U_{zero} value is rounded to the next upper natural number. The data matrix size is then N_{rg} by $P + 2U_{zero}$. If the illuminated scene X_g is equal or smaller than X_s then U_{zero} is equal to zero and no additional zeropadding in azimuth direction is necessary.

6.7 Verification of Sliding Spotlight SAR Processing Algorithms

A simulation has been developed to generate the sliding spotlight raw data. The simulation provides the two dimensional raw data for a defined scenario. The position of simulated targets is given in Cartesian coordinates as azimuth and range position.

A simulated raw data set was processed with presented algorithms in order to verify their focusing properties. Fig. 6.16 presents the scenario used for generating the raw data set. The scenario consists of 9 point targets. The first point target, referred with number 1, is positioned in the center of the scene in range and azimuth. Eight point targets numbered from 2 to 9 are positioned at the perimeter of a rectangle with a side length of two times 2.6km as shown in Fig. 6.16. The raw data was simulated with a flight altitude of 500km over the ground level. The incidence angle was set to 36.11° and the parameter for the sliding spotlight A equal to 0.3 was chosen. An oversampling of 1.25 for the sampling in azimuth direction and 1.3 for the sampling in range for raw data was used. The raw data was generated using the same scenario for a resolution of 0.5 m and 0.3 m in both range and azimuth direction. Parameters for both simulations are collected in Table 6.2. It is to be noted that no windowing was applied for the resulting radar image, hence the resolution estimated from simulation results is slightly smaller and is accordingly 0.44 m and 0.27 m.

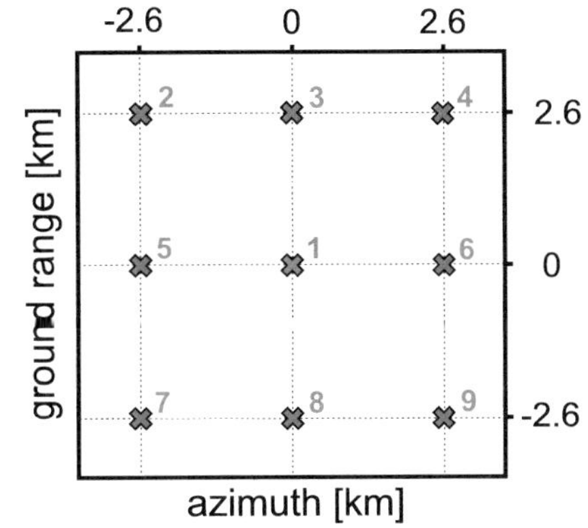

Figure 6.16: Test scenario with nine point targets.

Parameter	0.5 m res. simulation	0.3 m res. simulation
Orbit height	500 km	500 km
Incidence angle	36.11°	36.11°
Bandwidth	300 MHz	500 MHz
Effective antenna length	3.34 m	2 m
Chirp duration	70 μs	70 μs

Table 6.2: Simulation parameters.

6.7.1 Focusing Properties of the Sliding Spotlight Algorithms

Result for simulations of the test scenario for both processing algorithms described in Section 6.4 are presented. The chirp scaling algorithm and the omega-K algorithm for the sliding spotlight mode are verified for two resolutions 0.5m and 0.3m. Two point targets are presented in detail as examples for each simulation: point 1 from the center of the scene and point 9 positioned at −2.6 km in range and at 2.6 km in azimuth.

Chirp Scaling Algorithm for the Sliding Spotlight Mode for Airborne SAR

Results for the chirp scaling algorithm are presented in this paragraph. The test data for 0.5 m resolution are presented first. The Table 6.3 shows results for all simulated targets in the test scenario. The resolution of all nine simulated point targets is consistent with theoretical values calculated from simulation parameters and is about 0.44 m in range direction and 0.44 m in azimuth direction. Results for: points 1 and 9 are presented in more detail in Fig. 6.17 and Fig. 6.18 accordingly.

target no.	range pos. [km]	azimuth pos. [km]	range res. [m]	azimuth res. [m]
1	0	0	0.438	0.442
2	2.69	-2.69	0.440	0.441
3	2.69	0	0.440	0.440
4	2.69	2.69	0.439	0.442
5	0	-2.69	0.438	0.443
6	0	2.69	0.440	0.443
7	-2.69	-2.69	0.439	0.440
8	-2.69	0	0.440	0.440
9	-2.69	2.69	0.439	0.441

Table 6.3: Simulation of the airborne chirp scaling algorithm for 0.5 m resolution.

Fig. 6.17 shows results of a simulation with a resolution of 0.5m in both range and azimuth for the point target 1 positioned in the middle of the illuminated scene. Fig. 6.17(a) shows the magnitude of the compressed point target 1 as a two dimensional image in the time-space domain. The range and azimuth oversampling of the image was increased to improve the visibility of targets features. Cuts in range direction and azimuth direction through the center of the target's response are presented in Fig. 6.17(c) and Fig. 6.17(d) accordingly. Fig. 6.17(b) shows the phase of the point target's spectrum as a two dimensional image in range frequency and azimuth frequency. Fig. 6.17(e) and Fig. 6.17(f) show the cut in range frequency and azimuth frequency through the center of the phase spectrum accordingly. All point targets presented in following are described in the same way.

Characteristic for an correctly compressed point target is symmetry in range and in azimuth in time-space domain. A linear phase spectrum characteristic is another feature of a correctly compressed and distortion free point target. In Fig. 6.17 the target's response is symmetrical in both range and azimuth direction. The phase characteristic is linear and no distortions are visible. The resolution is 0.44 m in range and also 0.44 m in azimuth. This result corresponds to the resolution calculated for simulation parameters which is around 0.445 m.

Results for the point target 9 are presented in Fig. 6.18. This target is positioned at the edge of the illuminated area. The resulting resolution is 0.44 m for range and 0.44 m for azimuth direction which corresponds to the theoretical resolution. No defocussing or distortions are visible in space nor in frequency domain. The chirp scaling algorithm works very well for the airborne sliding spotlight mode with resolution of 0.5 m.

The simulation was repeated for the chirp scaling algorithm for a resolution of 0.3 m. The same scenario with nine point target was used for this purpose. Table 6.4 presents results for range and azimuth resolution for all nine simulated targets. The resolution for all simulated targets corresponds well with expected theoretical value. The point 1 and 9 are presented in more detail.

Results for the point target 1 are presented in Fig. 6.19. The resolution of this target is 0.27 m for the range direction and 0.27 m for the azimuth direction. The phase characteristic in rage and azimuth frequency is linear. The point target is symmetrical in azimuth direction. A slight difference in high of side lobes in range direction of the point target's is visible in these figures. This effect can also be seen in the two dimension point targets image in Fig. 6.19(a).

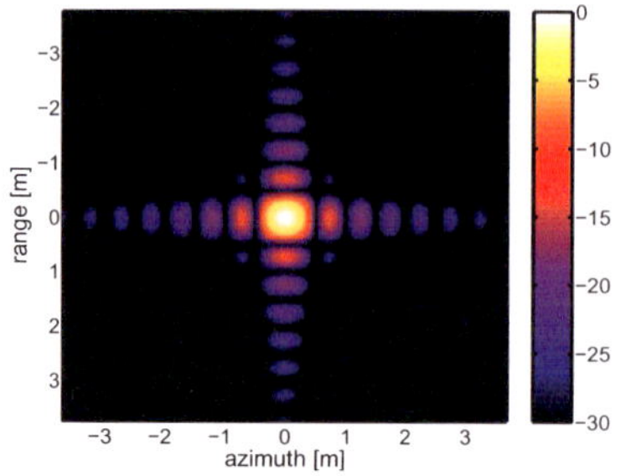

(a) two dimensional radar image

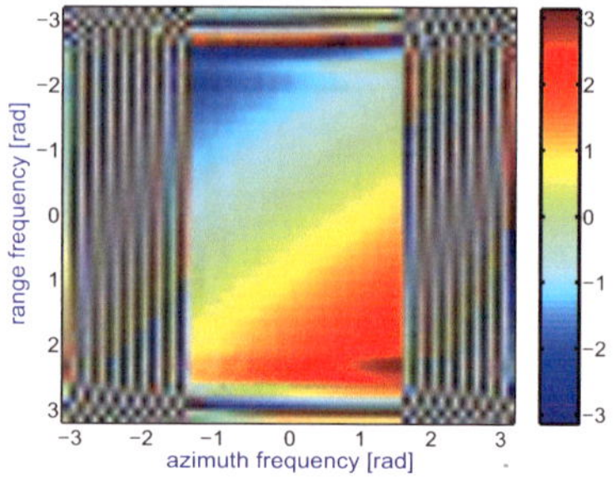

(b) two dimensional phase characteristic

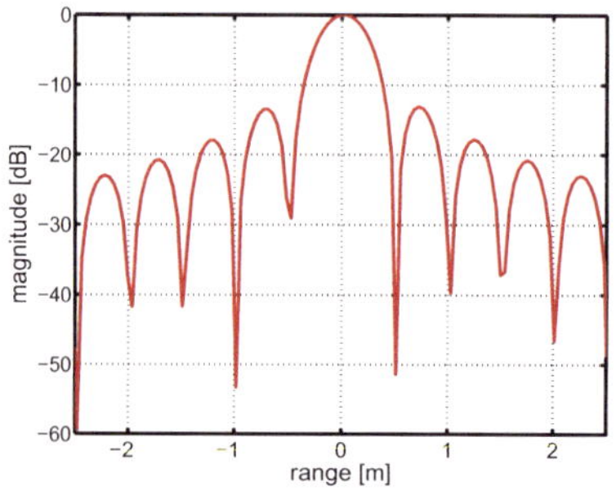

(c) radar image range profile

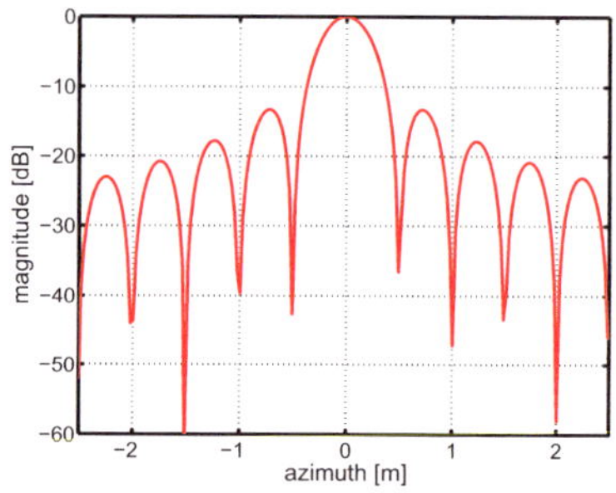

(d) radar image azimuth profile

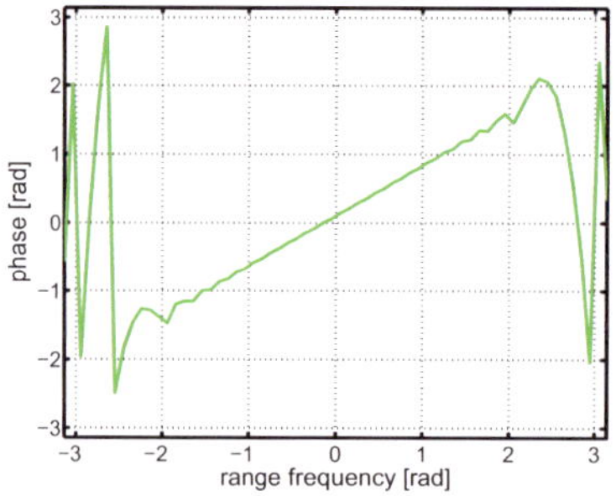

(e) phase characteristic cut in range frequency

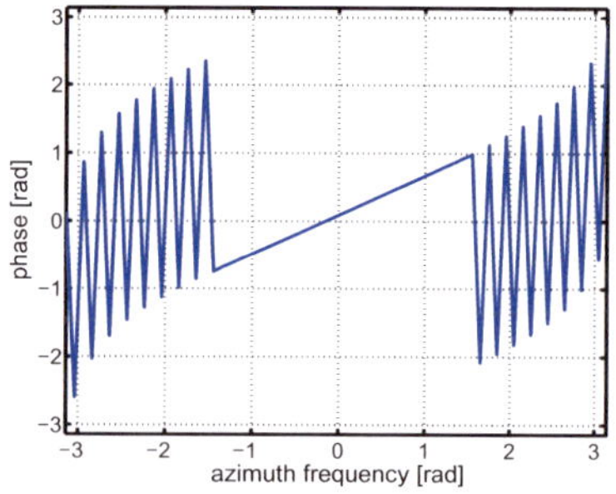

(f) phase characteristic cut in azimuth frequency

Figure 6.17: Simulation results for point target 1 with 0.5 m resolution compressed with the chirp scaling algorithm.

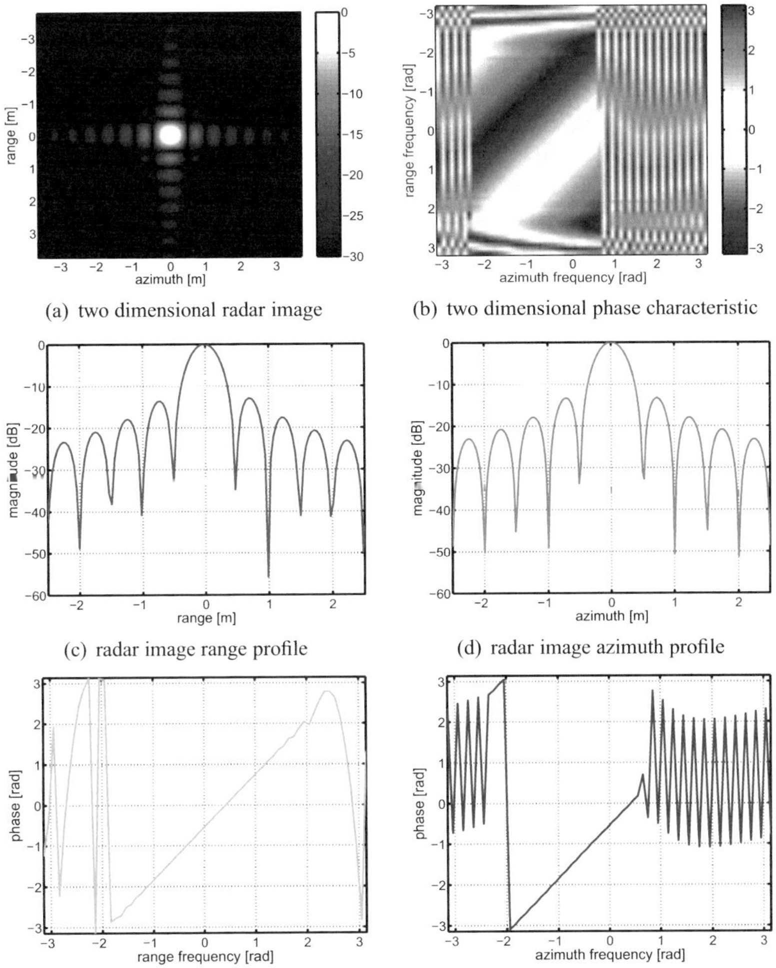

(a) two dimensional radar image

(b) two dimensional phase characteristic

(c) radar image range profile

(d) radar image azimuth profile

(e) phase characteristic cut in range frequency

(f) phase characteristic cut in azimuth frequency

Figure 6.18: Simulation results for point target 9 with 0.5 m resolution compressed with the chirp scaling algorithm.

target no.	range pos. [km]	azimuth pos. [km]	range res. [m]	azimuth res. [m]
1	0	0	0.266	0.272
2	2.69	-2.69	0.268	0.272
3	2.69	0	0.265	0.269
4	2.69	2.69	0.274	0.276
5	0	-2.69	0.269	0.275
6	0	2.69	0.274	0.279
7	-2.69	-2.69	0.270	0.278
8	-2.69	0	0.265	0.275
9	-2.69	2.69	0.275	0.281

Table 6.4: Simulation of the airborne chirp scaling algorithm for 0.3 m resolution.

Results for the point target 9 are presented in Fig. 6.20. The phase characteristic is linear in both range and azimuth frequency. The point target in space dimension shows slight defocussing in the range direction. The point target is symmetrical in azimuth direction. The resolution for this point target compressed with the chirp scaling algorithm is 0.28 m for the range direction and 0.28 m for the azimuth direction. The slight increase of the resolution is acceptable for a SAR processing algorithm.

Omega-K Algorithm for the Sliding Spotlight Mode for Airborne SAR

The raw data from simulation with 0.5 m and 0.3 m resolution were processed with a second algorithm proposed for sliding spotlight mode: the omega-K algorithm. First presented are results for simulation with 0.5 m resolution. Achieved resolution values are presented in Table 6.5. Results correspond well with the expected theoretical values of 0.44 m resolution in range and in azimuth. The point target 1 and the point target 9 are presented in more detail.

The point target 1 is presented in Fig. 6.22. The point target in space domain is symmetrical in range and azimuth direction. The phase characteristic of the target is linear in both range and azimuth frequency. The resolution of the

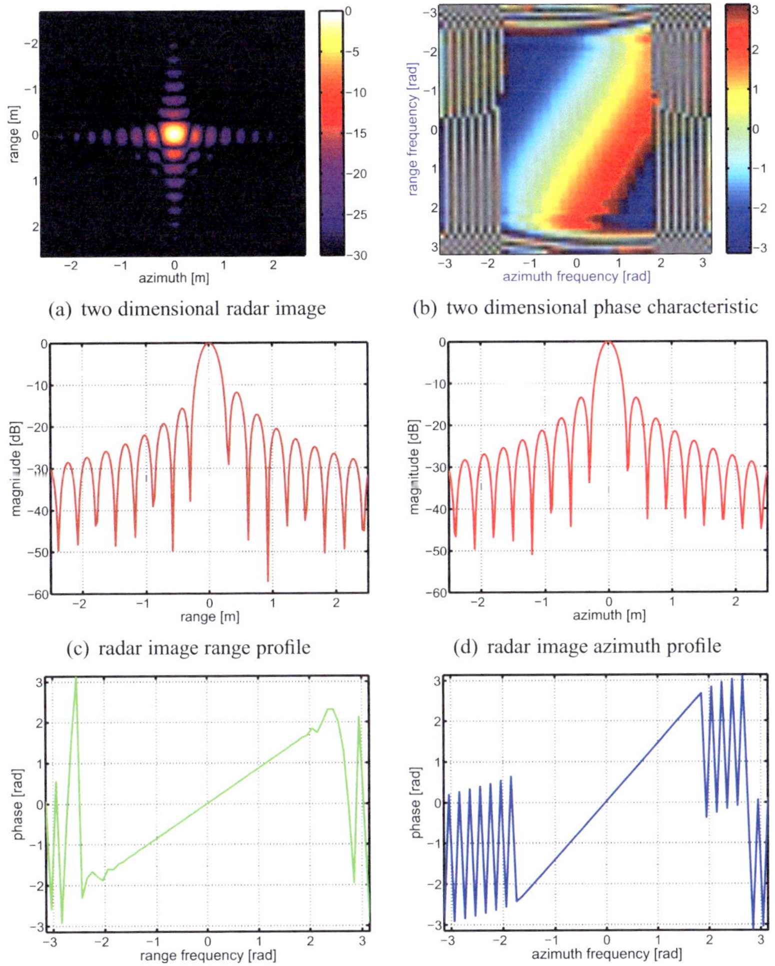

(a) two dimensional radar image (b) two dimensional phase characteristic

(c) radar image range profile (d) radar image azimuth profile

(e) phase characteristic cut in range frequency (f) phase characteristic cut in azimuth frequency

Figure 6.19: Simulation results for point target 1 with 0.3 m resolution compressed with the chirp scaling algorithm.

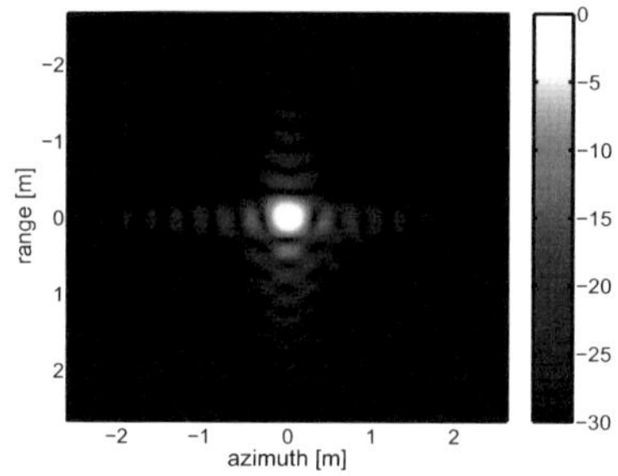

(a) two dimensional radar image

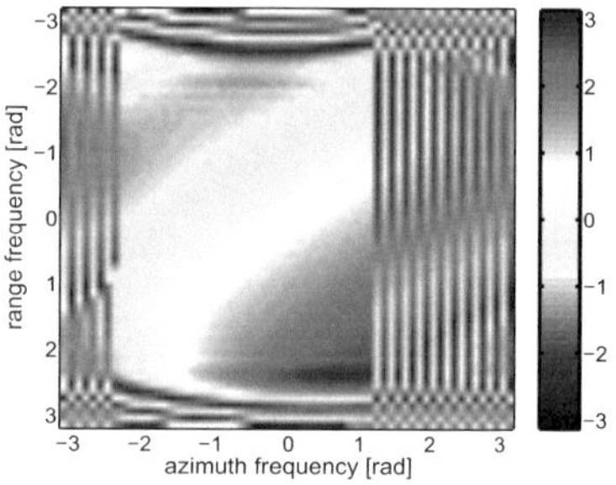

(b) two dimensional phase characteristic

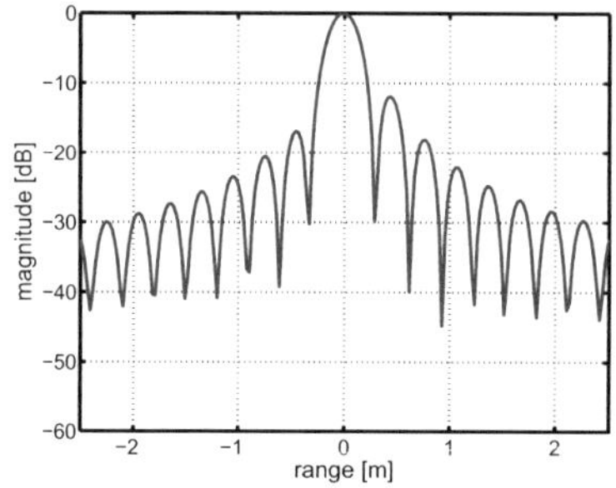

(c) radar image range profile

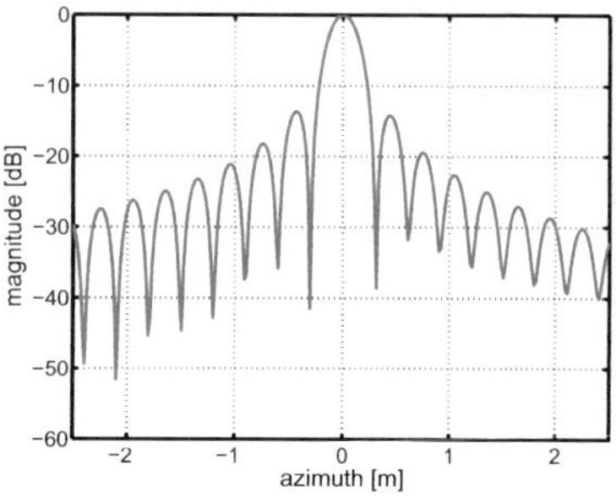

(d) radar image azimuth profile

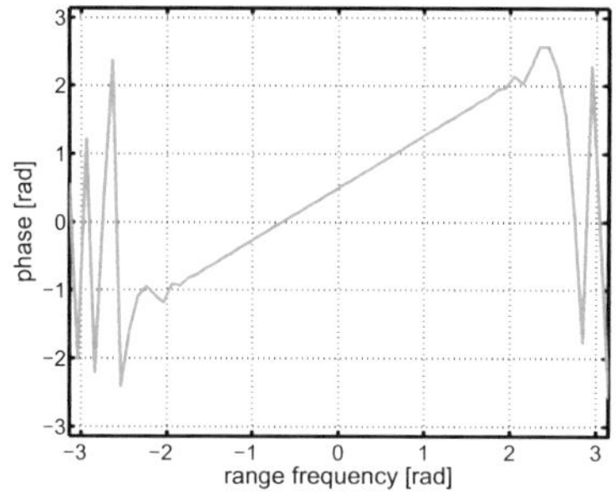

(e) phase characteristic cut in range frequency

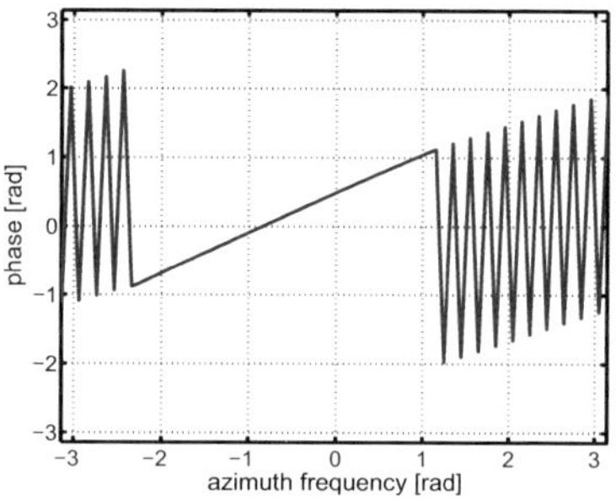

(f) phase characteristic cut in azimuth frequency

Figure 6.20: Simulation results for point target 9 with 0.3 m resolution compressed with the chirp scaling algorithm.

target no.	range pos. [km]	azimuth pos. [km]	range res. [m]	azimuth res. [m]
1	0	0	0.438	0.442
2	2.69	-2.69	0.440	0.441
3	2.69	0	0.440	0.449
4	2.69	2.69	0.440	0.440
5	0	-2.69	0.438	0.442
6	0	2.69	0.438	0.441
7	-2.69	-2.69	0.439	0.441
8	-2.69	0	0.439	0.443
9	-2.69	2.69	0.439	0.441

Table 6.5: Simulation of the airborne omega-K algorithm for 0.5m resolution.

target 1 resulting from simulations is 0.44 m in range and 0.44 m in azimuth. The point target 9 positioned at the edge of the illuminated scene is presented in Fig. 6.22. The target in space domain is symmetrical in both directions. The phase characteristic is linear in range and azimuth frequency. The resolution calculated for the simulation in both range and azimuth direction is equal to 0.44 m. The omega-K algorithm provides a very good compressed image for the sliding spotlight mode SAR with 0.5 m resolution. Results of the simulation for all point targets from the scenario are consistent with the theoretical values calculated for simulation parameters.

The nine point target scenario was simulated with a resolution of 0.3 m and processed with the omega-K algorithm. Results for the achieved resolution for all targets are presented in Table 6.6. All results correspond well with the theoretical value. It is to be noted that the values for achieved resolution are slightly better than for the data compressed with the chirp scaling algorithm. The point target 1 and 9 are presented in more detail.

The point target 1 is presented in Fig. 6.23. The resolution value for this target is about 0.26 m for the range and azimuth direction. The target in space domain is symmetrical in both range and azimuth direction. No distortions or defocussing is visible. The phase characteristic is linear in both range and azimuth frequency. The point 9 is presented in Fig. 6.24. The resolution calculated from the simulation of this point target is 0.26 m for range direction and 0.26 m for the azimuth direction. This corresponds with the expected

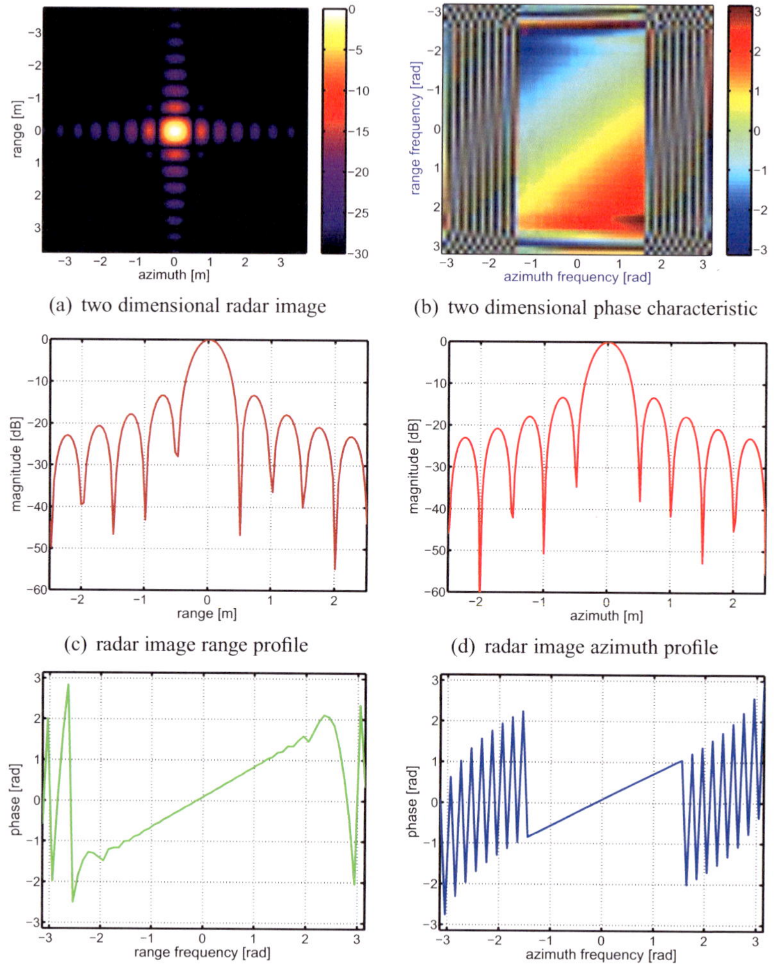

(a) two dimensional radar image

(b) two dimensional phase characteristic

(c) radar image range profile

(d) radar image azimuth profile

(e) phase characteristic cut in range frequency

(f) phase characteristic cut in azimuth frequency

Figure 6.21: Simulation results for point target 1 with 0.5 m resolution compressed with the omega-K algorithm.

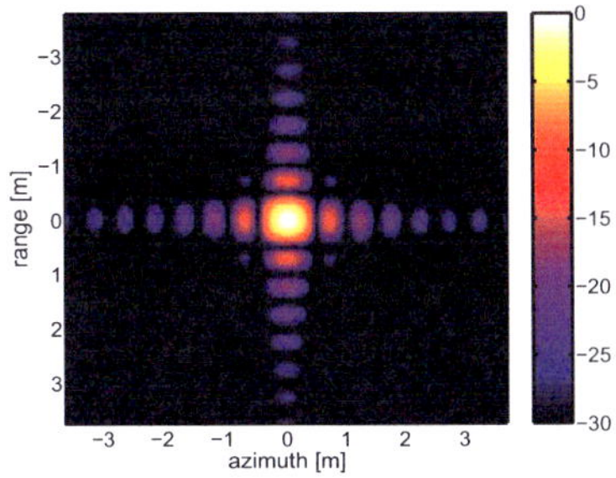

(a) two dimensional radar image

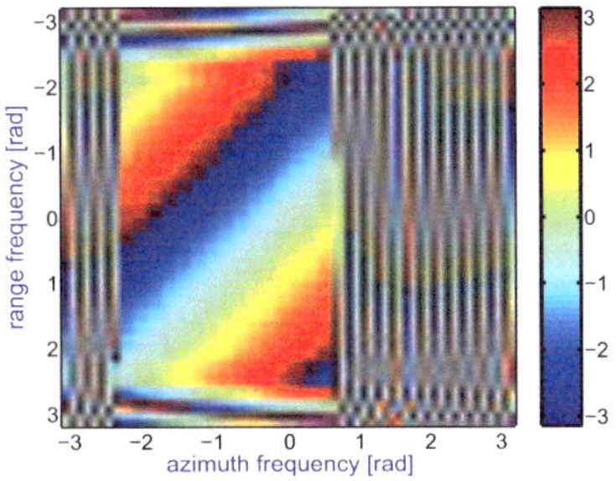

(b) two dimensional phase characteristic

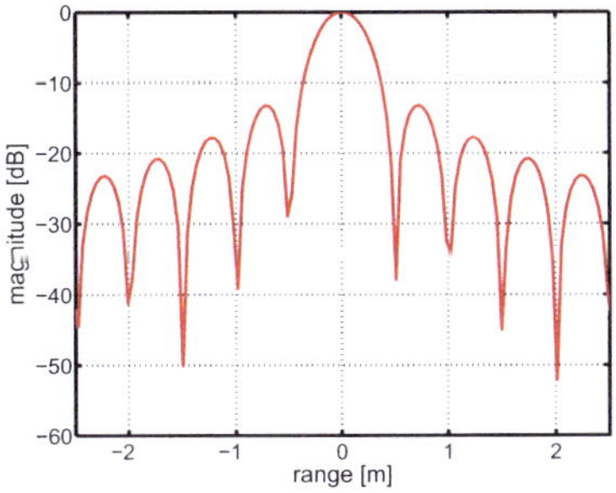

(c) radar image range profile

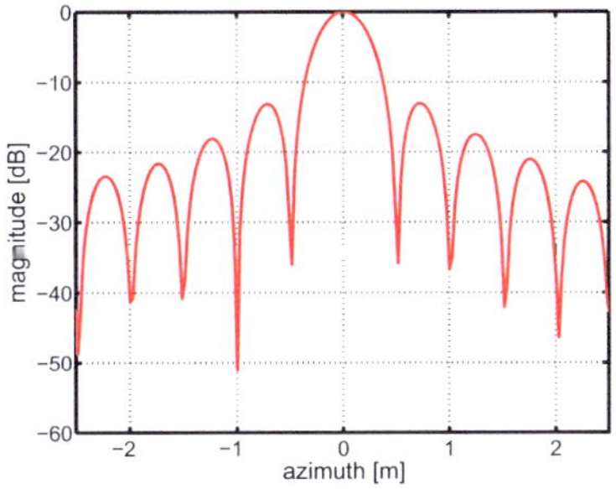

(d) radar image azimuth profile

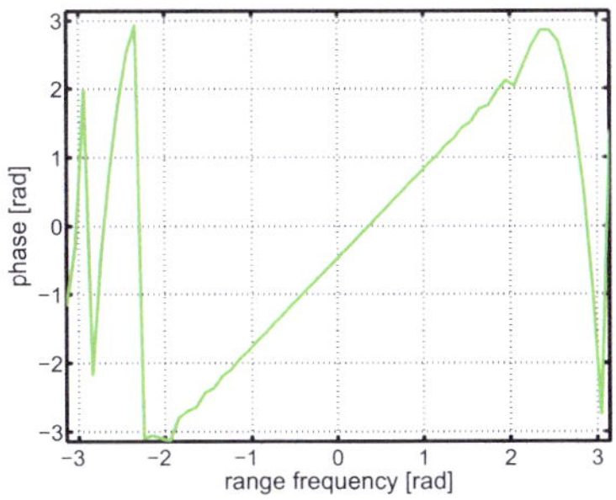

(e) phase characteristic cut in range frequency

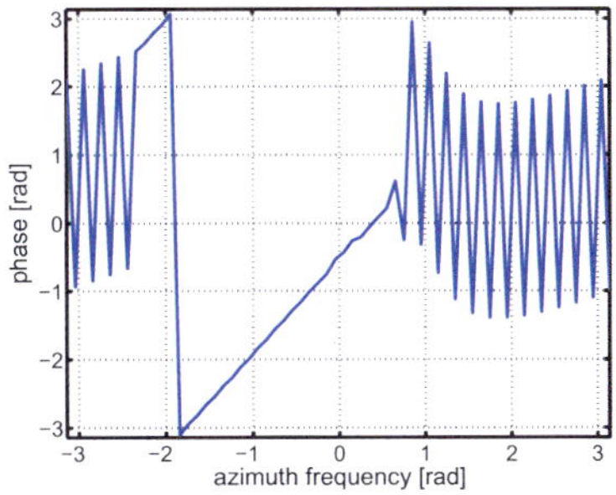

(f) phase characteristic cut in azimuth frequency

Figure 6.22: Simulation results for point target 9 with 0.5 m resolution compressed with the omega-K algorithm.

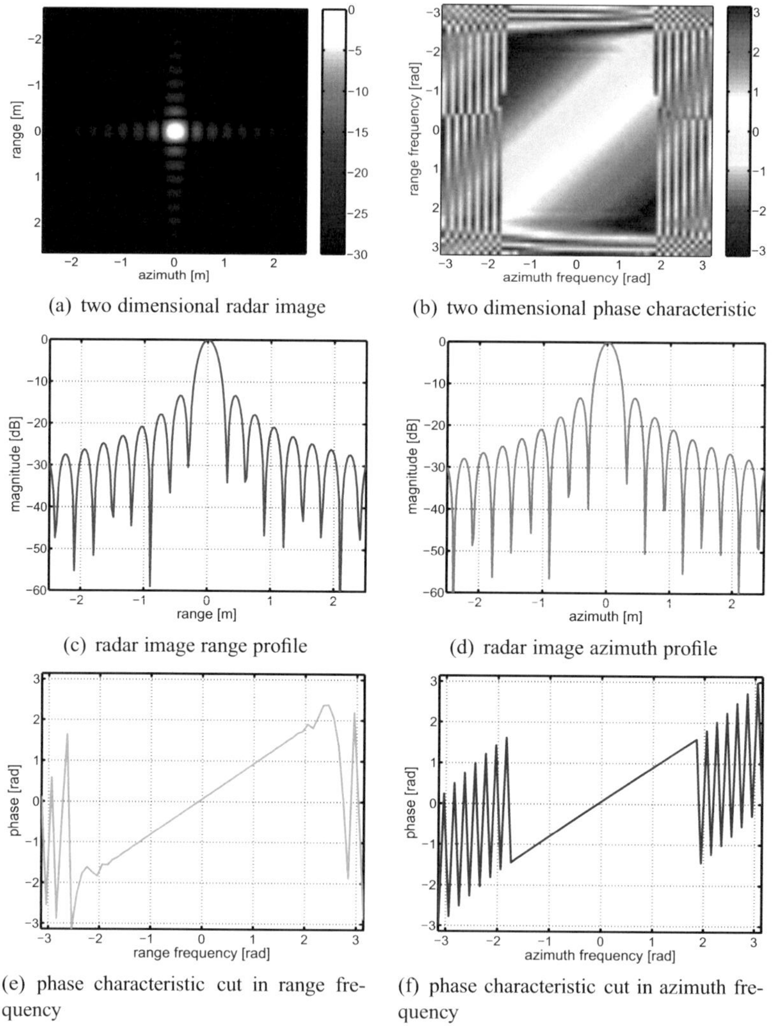

(a) two dimensional radar image

(b) two dimensional phase characteristic

(c) radar image range profile

(d) radar image azimuth profile

(e) phase characteristic cut in range frequency

(f) phase characteristic cut in azimuth frequency

Figure 6.23: Simulation results for point target 1 with 0.3 m resolution compressed with the omega-K algorithm.

target no.	range pos. [km]	azimuth pos. [km]	range res. [m]	azimuth res. [m]
1	0	0	0.263	0.264
2	2.69	-2.69	0.263	0.265
3	2.69	0	0.263	0.266
4	2.69	2.69	0.263	0.266
5	0	-2.69	0.263	0.264
6	0	2.69	0.263	0.265
7	-2.69	-2.69	0.263	0.264
8	-2.69	0	0.263	0.264
9	-2.69	2.69	0.263	0.265

Table 6.6: Simulation of the airborne omega-K algorithm for 0.3m resolution.

values. The point target in space domain is symmetrical in range and in azimuth direction. The phase characteristic of the target positioned at the edge of the illuminated scene is also linear in both range and azimuth frequency. The omega-K algorithm is applicable for sliding spotlight mode with 0.3 m resolution. All target in the scene are focused correctly.

6.8 Summary

An airborne sliding spotlight mode SAR was described in this chapter. The advantage of this mode is higher azimuth resolution than in strip-map mode and longer illuminated swath than in spotlight mode. The geometry of this mode and its influence on the received raw data was considered. The steering of the radar antenna during the data acquisition results in undersampling of the collected radar raw data. This can be resolved with an adequate signal processing. A proposal for two new SAR algorithms for processing airborne sliding spotlight mode raw data has been presented. An azimuth filtering of the radar raw data has been applied to the two chosen processing algorithms: the chirp scaling algorithm and the omega-K algorithm. The azimuth filter applied in the pre-processing step on the raw data requires a compensation in the compression algorithm. An algorithm for the compression of the antenna pattern weighting in SAR data has been presented. Furthermore a solution

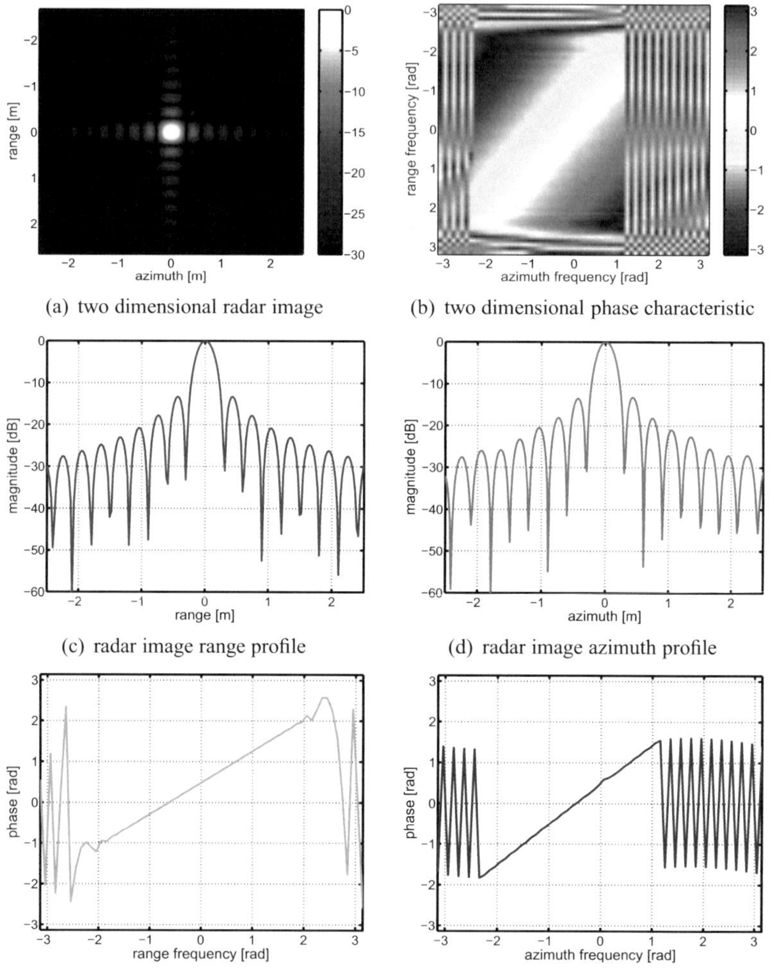

(a) two dimensional radar image

(b) two dimensional phase characteristic

(c) radar image range profile

(d) radar image azimuth profile

(e) phase characteristic cut in range frequency

(f) phase characteristic cut in azimuth frequency

Figure 6.24: Simulation results for point target 9 with 0.3 m resolution compressed with the omega-K algorithm.

for the wrap around of radar image for a specific configuration of system parameters was described. The performance of the compression algorithms have been evaluated with a simulation. The test scenario consisting of nine point targets positioned in the middle and on edges of the illuminated scene have been used for generation of test data. A resolution of 0.5 m and 0.3 m has been simulated. Both proposed algorithms allow processing of the highly resolved radar data with no significant resolution loss or performance loss due to algorithm inaccuracy. The chirp scaling algorithm is easier to implement as it consists only of matrix multiplications. The omega-K algorithm is more suited for high resolution radar data. This difference is also visible in achieved simulation results. The simulation results for the 0.3 m resolution data compressed with the chirp scaling algorithm shows some minor degradation due to inaccuracy of the processing algorithm. The degradation of the resulting radar image is still negligible for a SAR system. The omega-K algorithm shows a very good performance for both simulated resolutions. This algorithm requires more computing power. It provides however better accuracy for highly resolved radar data. The Omega-K algorithm is suggested for resolutions better than 0.3 m. The chirp scaling algorithm is providing same accuracy for 0.5 m simulation with less computing power. This algorithm would be advisable for those resolutions.

7 Processing of Spaceborne SAR Sliding Spotlight Mode

A detailed description of the sliding spotlight mode and processing algorithms for this mode have been given in the previous chapter. Chapter 6 considers an airborne SAR system where the radar is mounted on a platform such as e.g. an airplane or a helicopter. In case of the airborne SAR, as described in Chapter 6, the trajectory of the radar platform can be assumed to be a straight line. This approximation of a flat Earth is fully sufficient for processing algorithms for airborne SAR, where the radar is moving in a relatively low altitude above the illuminated ground. The approximation of a flat Earth is no longer valid for a spaceborne SAR system where the radar platform is flying few hundreds kilometers over the Earth surface. A more complex model of the satellite trajectory is necessary for SAR compression algorithms. The non straight trajectory of the radar platform results in a different Doppler history and different RCM for a target at a specific range. To achieve a fully compressed radar image both effects must be accordingly compensated.

The main focus in this chapter is put on processing algorithms for the sliding spotlight mode for the spaceborne SAR. The difference in the airborne and the spaceborne SAR geometry and its influence on the radar raw data are described. The chirp scaling algorithm and the omega-K algorithm are considered for compression of the SAR raw data. Both algorithms described already in the previous chapter are developed for an airborne SAR with the assumption of a straight line trajectory. The definition of parameters describing the sliding spotlight mode needs to be adjusted for a radar platform flying on a non-straight trajectory. The compression algorithm must also correct the difference between straight and the non-straight trajectory. Necessary changes are presented and implemented into both SAR processing algorithms the chirp scaling algorithm and the omega-K algorithm or the sliding spotlight mode. A very simple model of the satellite orbit and the Earth surface is used at first. The orbit of the satellite is modeled as an ideal circle and the Earth is assumed an ideal round sphere. SAR processing algorithms are adjusted

to compress raw data simulated with this trajectory and illuminated scenario model. The spaceborne chirp scaling and the omega-K algorithm for sliding spotlight mode are verified through adequate simulations presented in following sections.
Future SAR systems will provide images with very high resolution over long ground swath. This results in a much longer synthetic aperture and a much longer flight trajectory. The increased resolution requires also a more precise approximation of the trajectory of the radar platform. A much longer synthetic aperture requires more accurate processing and more precise modeling of the satellite trajectory. Following sections consider the possible compensation of an arbitrary satellite trajectory. The possible compensation of a real platform trajectory is described.

7.1 Spaceborne Sliding Spotlight Mode

The flat Earth model for the geometry of a SAR system was considered in the previous chapter for processing algorithms and used for simulations of the raw data. This model is fully sufficient for an airborne radar, where it can be assumed that the radar platform is moving on a straight line trajectory and the curvature of the Earth seen from a typical flight altitude of an airplane can be neglected. In the spaceborne radar system the curvature of the Earth needs to be considered as well as the non-straight radar platform trajectory. Fig. 7.1 presents the geometry of the spaceborne SAR sliding spotlight mode. Fig. 7.1(a) shows a three dimensional representation of the spaceborne sliding spotlight mode geometry. The Cartesian coordinate system has its origin in the center of the Earth sphere, which is modeled as an ideal round sphere. It is to be noted that in the spaceborne case the y-direction in the Cartesian coordinate system corresponds no longer to the azimuth direction. $\vec{u}$ is still corresponding to the trajectory of the radar platform in the flight direction as it is defined now as a function of time t and unit base vectors $\vec{e_x}$, $\vec{e_y}$ and $\vec{e_x}$.

$$\vec{u}(t) = a_x(t)\vec{e_x} + a_y(t)\vec{e_y} + a_z(t)\vec{e_z} \tag{7.1}$$

Where $a_x(t)$, $a_y(t)$ and $a_z(t)$ are coefficients for the x-, y- and z-direction accordingly. The direction given by the vector $\vec{u}(t)$ is described further as azimuth direction of the radar system.
The azimuth direction in case of spaceborne SAR is a part of a circle with a diameter of $R_{Earth} + h$ and the center in the center of the Earth. R_{Earth} is the

mean Earth radius[1] and h is the radius of the radar's orbit. Fig. 7.1(b) shows a cut in the flight direction of the geometry of the spaceborne sliding spotlight mode. The black point in Fig. 7.1(b) marks the center of the Earth sphere. The red point marks the steering point for the sliding spotlight mode. The satellite platform is moving on the orbit with the circular velocity ω_s and the antenna footprint is moving on the Earth's surface with the circular velocity ω_f.

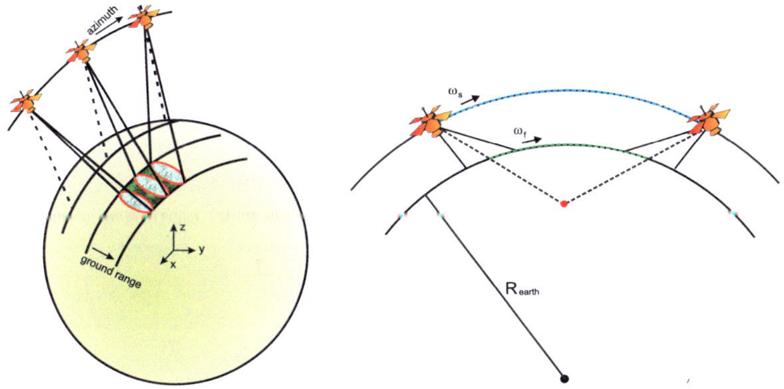

(a) a three dimensional view (b) cut through the azimuth direction

Figure 7.1: The sliding spotlight mode for satellite SAR.

Fig. 7.2(a) shows a cut in flight direction of the spaceborne SAR strip-map mode geometry. It is to be noted that in case of the spaceborne strip-map mode all beams of the radar antenna for different positions of the radar platform go through the same point, which can be considered as the steering point. The steering point in the case of the airborne strip-map mode is positioned in infinity. The lines of sight of different positions in the synthetic aperture of the radar antenna do not cross for the airborne case. The curvature of the flight trajectory for the spaceborne case causes lines of sight of the radar antenna to cross in the steering point positioned in the center of the Earth sphere.

Fig. 7.2(b) shows the spotlight mode geometry of the spaceborne SAR. The

[1] The mean Earth radius is equal to 6378.160 km [12].

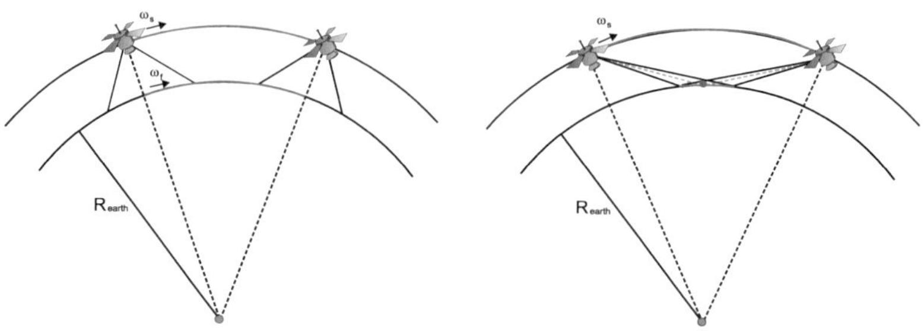

(a) the strip-map mode (b) the spotlight mode

Figure 7.2: The cut in azimuth direction in spaceborne SAR geometry for the strip-map mode and the spotlight mode.

radar antenna beams on various positions are crossing in a point on the Earth's surface. It is to be noted that the synthetic aperture for a target is longer in the spaceborne SAR case than it would be in the airborne case assuming the same parameters of radar system. This difference results from the curved flight path of the radar platform.

Parameters describing the sliding spotlight mode in the Section 6.1 have been defined for an airborne SAR. The parameter A has been defined as the ratio of the distance from the steering point to the ground and the distance between the radar antenna and the ground. The second definition of the parameter A is the ratio of the velocity of the radar antenna beam footprint on the ground to the velocity of the radar platform. The parameter A in the airborne case has the nice property of describing also the two limiting cases of the sliding spotlight mode: the strip-map mode with A equal to zero and the spotlight mode with A equal to one. This definition of the parameter A is not applicable in case of the spaceborne SAR. It can be clearly seen in Fig. 7.1(b), Fig.7.2(a) and Fig. 7.2(b).

A new definition of the parameter A for the spaceborne SAR sliding spotlight mode is proposed in this work. The parameter A is defined as the ratio of the circular velocity of the radar antenna footprint on the ground to the circular velocity of the radar platform.

$$A_{spaceborne} = A = \frac{\omega_f}{\omega_s} \tag{7.2}$$

Where ω_f is the circular velocity of the antenna beam footprint on the ground and ω_s is the circular velocity of the satellite. These velocities can be calculated with the knowledge of the orbit's alititude, the mean Earth radius (for other planets or moons the adequate radius needs to be considered) and the steering angle of the antenna beam. In further the index $_{spaceborne}$ will be omitted. This new definition of the parameter A for the spaceborne SAR retains dependencies between parameters and equations. Through the definition with circular velocities the limiting cases of the sliding spotlight mode are also preserved. The parameter A is equal to one for the strip-map mode and equal to zero for the spotlight mode, same as in the airborne SAR system. With this definition of the parameter A for spaceborne SAR no additional modifications for the azimuth filter are necessary. The azimuth filter is to be calculated according to Eq. 6.18 as for the airborne case.

7.2 Correction for an Ideal Circular Orbit for SAR Processing

The chirp scaling processing algorithm [76] and omega-K algorithm have been developed under the assumption of flat Earth geometry approximation. This approximation is no longer valid for a satellite SAR, [75]. The defocussing caused by approximations in processing algorithms is significant, especially for the high resolution radar data. An ideal circular orbit of the satellite is assumed for the algorithm development in this section. The correction to such an orbit in the chirp scaling algorithm can be implemented through an additional parameter called velocity ratio v_s/v_f [66]. The synthetic aperture for a single point target in the case of a spaceborne SAR is longer than for the airborne case assuming the same system parameters. The length of the synthetic aperture corresponds to the illumination time of the target. The difference in illumination time between the airborne and the spaceborne SAR geometry results from the curvature of the platform flight trajectory, which causes a slight 'steering' of the radar antenna after the illuminated scene. The difference in the illumination time can be related to the difference between the velocity of the satellite and the velocity of the antenna beam footprint on the ground for the strip-map mode. The velocity ratio is defined here as the square root of the satellite velocity v_s to the velocity of the antenna beam footprint on the ground for the airborne case $v_{f/strip-map}$. The velocity ratio

v_s/v_f can be calculated according to Eq. (7.3), [75].

$$\begin{aligned} v_s/v_f &= \sqrt{\frac{v_s}{v_{f/strip-map}}} \\ &= \sqrt{\frac{R_{Earth}+h}{R_{Earth}\cos\left(\arcsin\left(\frac{r_s}{R_{Earth}+h}\sin(\varphi_i)\right)\right)}} \end{aligned} \tag{7.3}$$

Fig. 7.3 shows the geometry for the calculation of the velocity ratio. It is to be noted that the footprint velocity $v_{f/strip-map}$ relates here to the strip-map mode, which is also indicated by the index 'strip-map'. The velocity of the antenna beam footprint is different for each mode. φ_i is the incidence angle which can be calculated for the spaceborne case according to Eq. (7.4).

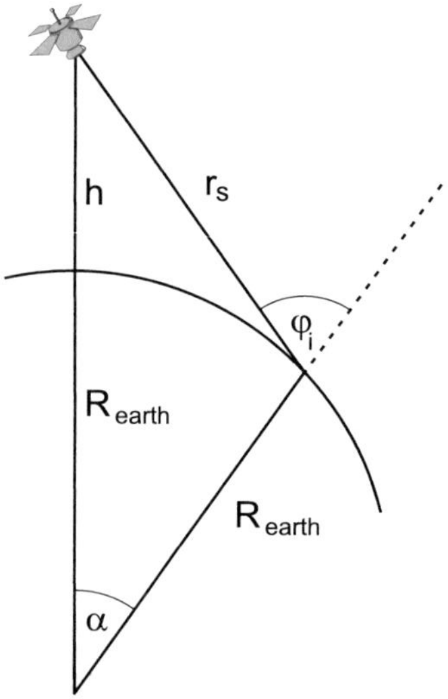

Figure 7.3: A cut in slant range direction of the spaceborne SAR geometry.

$$\varphi_i = \pi - \arccos\left(\frac{r_s^2 + R_{Earth}^2 - (r_s + R_{Earth})^2}{2r_sR_{Earth}}\right) \tag{7.4}$$

The angle α, see Fig. 7.3, is the angle between the center of the Earth, the radar antenna and the illuminated point target. The cosine of this angle can be calculated according to Eq. (7.5).

$$\begin{aligned}\cos(\alpha) &= \cos\left(\arcsin\left(\frac{r_s}{R_{Earth}+h}\sin(\varphi_i)\right)\right)\\ &= \frac{(R_{Earth}+h)^2+R_{Earth}^2-r_s^2}{2(R_{Earth}+h)R_{Earth}}\end{aligned} \tag{7.5}$$

Combining Eq. (7.4) with Eq. (7.5) and generalizing the equation for all ranges r we get an equation for the velocity ratio for all illuminated ranges, Eq. (7.6).

$$\begin{aligned}v_s/v_f(r) &= \sqrt{\frac{(R_{Earth}+h)}{R_{Earth}\frac{(R_{Earth}+h)^2+R_{Earth}^2-r^2}{2(R_{Earth}+h)R_{Earth}}}}\\ &= \sqrt{\frac{2(R_{Earth}+h)^2}{(R_{Earth}+h)^2+R_{Earth}^2-r^2}}\end{aligned} \tag{7.6}$$

It is to be noted that the velocity ratio $v_s/v_f(r)$ depends on range. For high resolution SAR images this dependence cannot be neglected in the processing. The ratio gives the dependence between the velocities of the satellite and the antenna beam footprint on the ground which corresponds to the ratio between the part of the radar trajectory the satellite traveled to the length of the illuminated swath in the same time period.

7.2.1 Chirp Scaling Algorithm for the Spaceborne Sliding Spotlight Mode

The implementation of necessary changes to the chirp scaling algorithm for the correction of an ideal orbit is described in this section. In SAR a longer illumination time of the target in the time domain corresponds with a higher maximal frequency of the target's response in the frequency domain. The correction of the ideal round orbit in the chirp scaling algorithm can be done with adequate substitution of the azimuth frequency f_{az}. The new azimuth frequency which is to be used for all calculations in the algorithm can be calculated according to Eq. (7.7).

$$f'_{az}(r) = v_s/v_f(r) f_{az} \tag{7.7}$$

It is to be noted that the velocity ratio as well as the new azimuth frequency depends on range. In this way Eq. (6.27) and Eq. (6.28) described in Section 6.4.2 for the chirp scaling algorithm become also range dependent. This concerns all exponential functions defined in Eq. (6.24), Eq. (6.29) and Eq. (6.31). There are no necessary changes in the azimuth filter. The Eq. (6.30) dose not require modifications. To ensure proper processing without aliasing a slight oversampling of the signal after the azimuth filter is applied. As the azimuth frequency in the radar data increases by the factor of the velocity ratio. The oversampling must be chosen not smaller than the velocity ratio. The oversampling of about 1.2 usually assumed for a SAR system is fully sufficient.
There are no further changes in the chirp scaling algorithm for the correction of the ideal circular orbit. The diagram of the chirp scaling algorithm for the spaceborne sliding spotlight mode is presented in Fig. 7.4(a). In Section 7.3 simulation results for a scene with nine point targets and various resolutions are presented.

7.2.2 Omega-K Algorithm for the Spaceborne Sliding Spotlight Mode

Corrections necessary for the omega-K algorithm in order to process a spaceborne SAR raw data are described in this section. As in the previous section an ideal round orbit over a spherical Earth is assumed. The azimuth frequency f_{az} is substituted with the new azimuth frequency $f'_{az}(r)$ as described in Eq. (7.7). The azimuth frequency is multiplied with the velocity ratio to compensate for the extended illumination time. To notice is that Eq. (6.33), Eq. (6.34) and Eq. (6.35) are now also range dependent.
A multiplication with an exponential function correcting for the Stolt interpolation calculated trajectory of the radar platform is necessary for the omega-K algorithm. The function is calculated according to Eq. (7.8).

$$\Phi_f(r, f_{az}) = \exp\left(-j2\pi\frac{\lambda f_{az}^2 r}{4v_s^2}\left(v_v/v_f(r)^2 - v_v/v_f(r_{ref})^2\right)\right) \quad (7.8)$$

The diagram of the spaceborne omega-K algorithm including necessary changes for spaceborne data can be seen in Fig. 7.4(b). The radar data is transformed into the range-Doppler domain after the Stolt interpolation and the correction of the azimuth filter. The necessary correction for a round orbit is applied on the compressed radar data in the range-Doppler domain. The data is then transformed into the time-space domain.

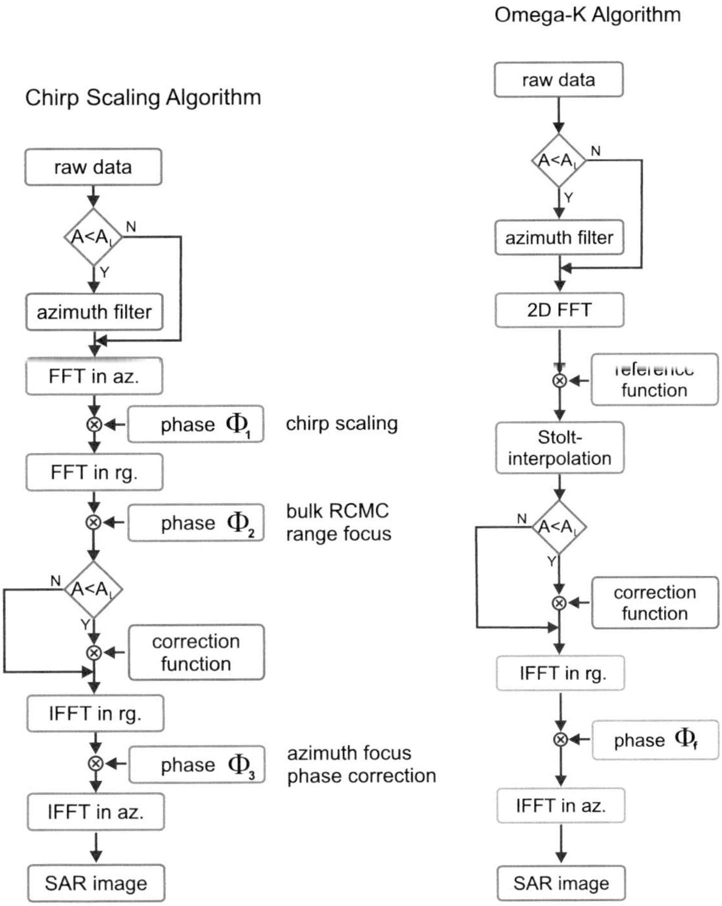

(a) chirp scaling algorithm (b) omega-K algorithm

Figure 7.4: Flow chart of the spaceborne processor for the sliding spotlight mode

7.3 Simulation Results for the Spaceborne Sliding Spotlight Mode

The simulation described in Chapter 6 was modified to generate radar raw data for a spaceborne SAR. The position of a target is given in the spheric coordinated. The trajectory of the satellite is modeled as a round orbit. Results of simulations for the spaceborne sliding spotlight SAR mode are presented in this section. Simulated radar raw data with various resolutions have been processed with the chirp scaling algorithm as described in Section 7.2.1 and with the omega-K algorithm as described in Section 7.2.2. The scene used for the simulation is similar as the one used in Chapter 6 for the simulation of the raw data for the airborne case, see Fig. 6.16. The difference is that point targets for the spaceborne SAR simulation are positioned on a surface of a sphere with the diameter of R_{Earth}. The position of point targets is: one in the center of the illuminated scene (the point target number 1) and eight targets positioned on edges of a rectangle as shown in Fig. 7.5(a). The rectangle edges calculated over the Earth's ground is twice the length of 2.69 km. The scene was simulated for a SAR system with 0.5 m and 0.3 m resolution. It is to be noted that no window is applied on the resulting radar image. Hence the resulting resolution is slightly smaller and is accordingly is equal to 0.44 m and 0.27 m in range and azimuth.

A slightly different scenario was used for an additional simulation with 0.2 m resolution (0.18 m without windowing) in both rage and azimuth. Only point targets 1 and 6 have been considered for this simulation. To distinguish the two scenarios the targets are named $1a$ and $6a$. Fig. 7.5(b) shows this configuration. The reason for limiting the number of point targets is the very high amount of the raw data necessary by this high resolution to be generated and processed. This simulated raw data was compressed only with the omega-K algorithm. The parameters for the SAR system for the simulation with 0.5 m, 0.3 m and 0.2 m resolution are presented in Table 7.1.

The result of simulations for various resolutions compressed with different SAR processing algorithms the chirp scaling and the omega-K algorithm are presented in following. A table with achieved resolution in range and azimuth for all nine point target is given for each simulation. The point targets 1 and 9 are presented in more detail. Six figures are describing the features of the target in the space and in the frequency domain. The configuration of presented

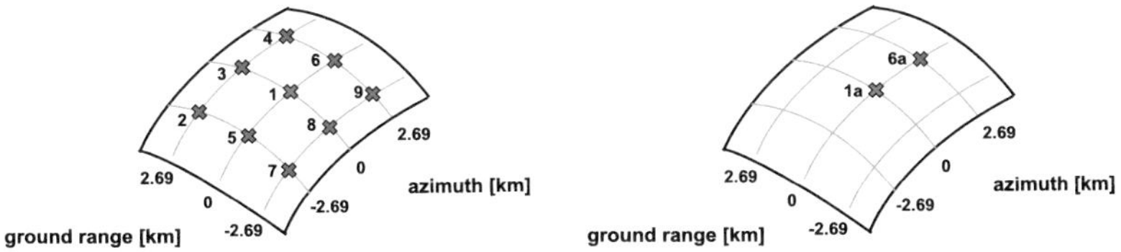

(a) nine point targets scenario (b) two point targets scenario

Figure 7.5: Test scenario with nine point targets for spaceborne SAR.

Parameter	0.5 m res. simulation	0.3 m res. simulation	0.2 m res. simulation
Orbit height	500 km	500 km	500 km
Incidence angle	36.11°	36.11°	36.11°
Bandwidth	300 MHz	500 MHz	750 MHz
Effective antenna length	3.34 m	2 m	1.34 m
Chirp duration	70 μs	70 μs	70 μs

Table 7.1: Simulation parameters for spaceborne SAR.

plots is the same for each of the point targets. Fig. (a) shows the two dimensional amplitude of the compressed point target in the space domain, Fig. (b) shows the two dimensional phase characteristic of the point target over range frequency and azimuth frequency, Fig. (c) and Fig. (d) present cuts of the amplitude of the compressed target in range direction and in azimuth direction accordingly, and Fig. (e) and Fig. (f) present cuts of the phase characteristic in range frequency and in azimuth frequency accordingly.

7.3.1 Chirp Scaling Algorithm for the Spaceborne Sliding Spotlight Mode

Results for a simulation with 0.5 m (0.44 m without windowing) resolution in both range and azimuth direction are presented in this section. The raw data was processed with the chirp scaling algorithm described in detail in Section 7.2.1. Fig. 7.6 and Fig. 7.7 show the results for the point target 1 and the point target 9 accordingly. For both point targets the achieved resolution is adequate to the theoretical value and is 0.44 m in range direction and 0.44 m in azimuth direction. Both point targets are symmetrical in space domain in range and in azimuth. The phase characteristic of these point targets' spectrum is linear. In Table 7.2 resolutions in range and in azimuth for all simulated targets are presented. The resolution is consistent for all targets with the theoretical value calculated from simulation parameters. The difference[2] between the 'worst' and the 'best' compressed point target in range direction is about 0.41% and in azimuth direction about 0.29%. The spread in the resolution in both range and azimuth direction for a SAR system is negligible.

target no.	range pos. [km]	azimuth pos. [km]	range res. [m]	azimuth res. [m]
1	0	0	0.439	0.441
2	2.69	-2.69	0.440	0.442
3	2.69	0	0.440	0.441
4	2.69	2.69	0.439	0.441
5	0	-2.69	0.439	0.441
6	0	2.69	0.438	0.440
7	-2.69	-2.69	0.440	0.442
8	-2.69	0	0.440	0.442
9	-2.69	2.69	0.440	0.442

Table 7.2: Simulation for spaceborne chirp scaling algorithm for 0.5 m resolution.

The same test scenario was used to simulate the raw data for a SAR system with 0.3 m resolution. Fig. 7.8 and Fig. 7.9 present results of the simulation for the point target 1 and 9 with resolution 0.3 m (0.26 m without windowing) in both range and azimuth direction. The raw data was processed with the spaceborne SAR chirp scaling algorithm for sliding spotlight mode. The resolution achieved in the simulation of the point target 1 is 0.27 m in range direction and 0.26 m in azimuth direction. The resolution resulting from the simulation for the point target 9 is 0.28 m in range direction and 0.27 m in

[2] Normalized to the 'best' compressed point target.

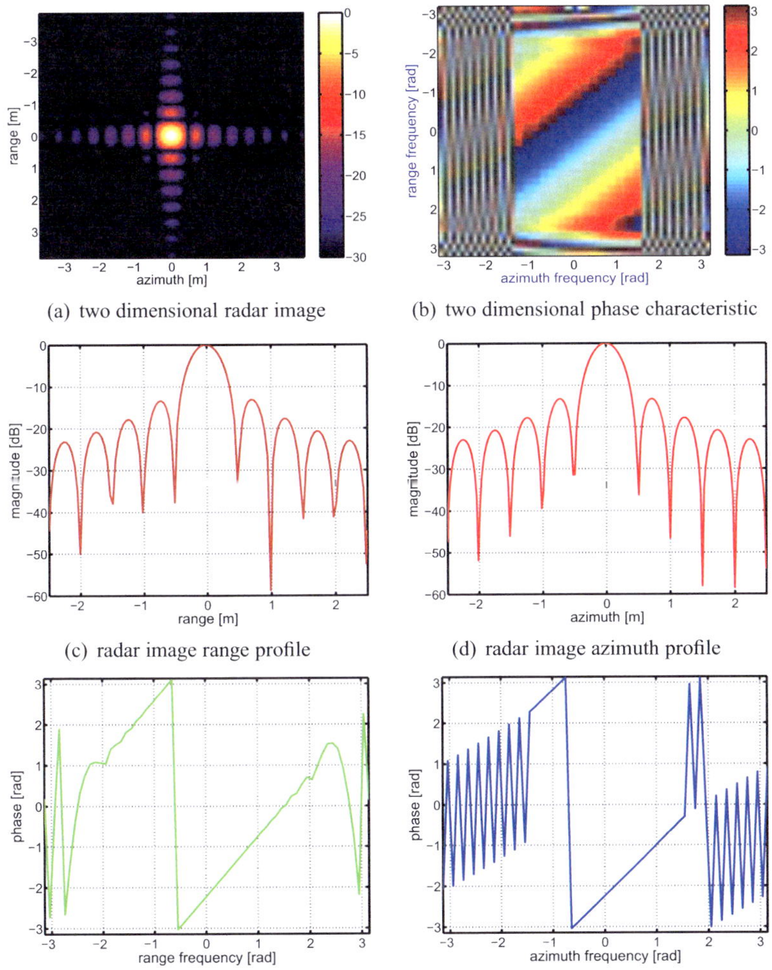

(a) two dimensional radar image

(b) two dimensional phase characteristic

(c) radar image range profile

(d) radar image azimuth profile

(e) phase characteristic cut in range frequency

(f) phase characteristic cut in azimuth frequency

Figure 7.6: Point target 1 with 0.5m resolution processed with the spaceborne chirp scaling algorithm.

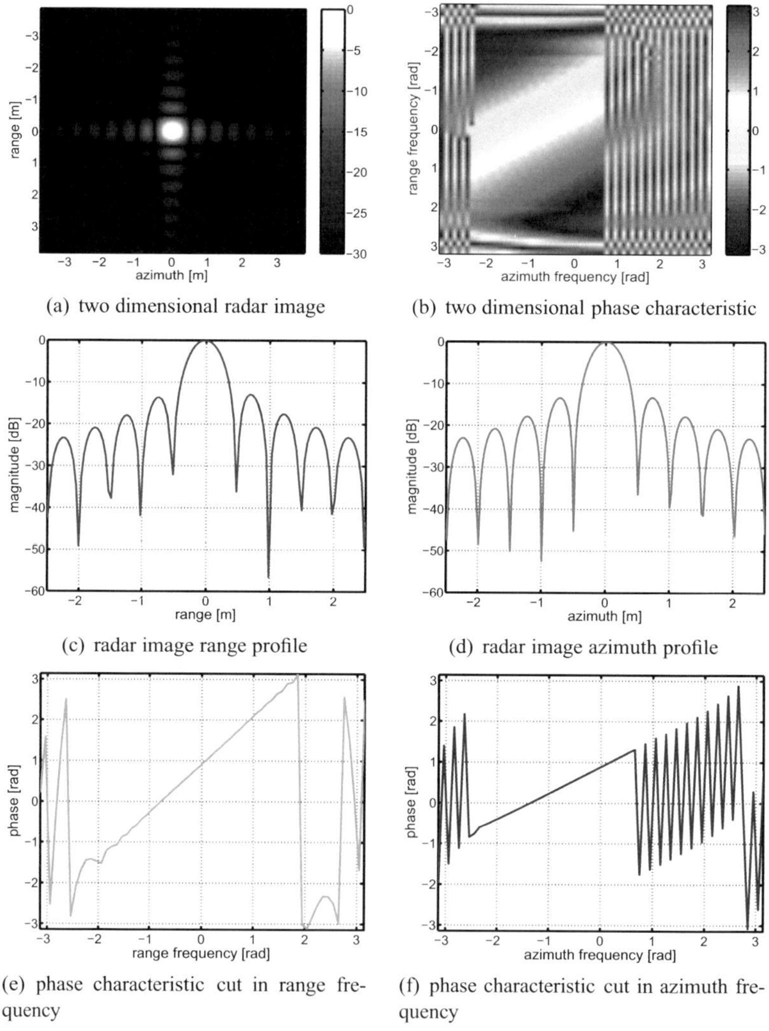

(a) two dimensional radar image

(b) two dimensional phase characteristic

(c) radar image range profile

(d) radar image azimuth profile

(e) phase characteristic cut in range frequency

(f) phase characteristic cut in azimuth frequency

Figure 7.7: Point target 9 with 0.5m resolution processed with the spaceborne chirp scaling algorithm.

azimuth direction. These results are slightly degraded compared with the expected theoretical value of 0.26 m in both range and azimuth direction. In both Fig. 7.8(c) and Fig. 7.9(c) a slight difference in the height of the first side lobes on the left and right of the main lobe is present. The two dimensional image of the point target in Fig. 7.8(a) shows also additional strong side lobes. In the phase characteristic presented in Fig. 7.8(c) and Fig. 7.9(c) is still relatively linear. Table 7.3 shows all results for range and azimuth direction resolution for this simulation. The difference between the achieved and the expected resolution value for all point targets is similar.

target no.	range pos. [km]	azimuth pos. [km]	range res. [m]	azimuth res. [m]
1	0	0	0.267	0.255
2	2.69	-2.69	0.270	0.260
3	2.69	0	0.266	0.256
4	2.69	2.69	0.272	0.262
5	0	-2.69	0.270	0.260
6	0	2.69	0.273	0.262
7	-2.69	-2.69	0.273	0.269
8	-2.69	0	0.266	0.265
9	-2.69	2.69	0.277	0.270

Table 7.3: Simulation for spaceborne chirp scaling algorithm for 0.3 m resolution.

The slight deformation of the phase characteristic for the simulation with 0.3 m resolution results in a certain decrease of the performance. The difference between the theoretically expected resolution and the worst achieved resolution from the simulation of the point target 9 is 3.6% for the range direction. The difference between the theoretical resolution and the worst achieved in simulation in azimuth direction is only 1.3%. Both values are acceptable errors for the processing of the SAR data. The decrease of the performance of the algorithm indicates that for higher resolutions or wider swaths a more precise algorithm is necessary.

The chirp scaling algorithm is a very good trade-off between maturity, simplicity, efficiency and accuracy, [12]. It is efficient and fast to implement as it requires only FFTs and phase multiplications. The algorithm has some disadvantages which become apparent under certain conditions. The algorithm is based on the approximation of the Taylor series expansion to estimate the range IFFT to transform the signal from the two dimensional frequency domain back into the range Doppler domain. This approximation is not adequate for high squint angles and for wide apertures. The approximation error becomes visible in Fig. 7.8 and Fig. 7.9. The chirp scaling algorithm is a good algorithm for processing the satellite SAR data of 0.5 m and 0.3 m resolution as shown in this section. For higher resolution a more precise algorithm is necessary.

7.3.2 Omega-K Algorithm for the Spaceborne Sliding Spotlight Mode

Results for simulations of the raw data compressed with the omega-K algorithm for 0.5 m, 0.3 m and 0.2 m resolution are presented in this section. The same test scenario consisting of nine point targets was used for 0.5 m and 0.3 m resolution simulation, see Fig. 7.5(a). A simplified two point scenario was used for simulation with 0.2 m resolution, see Fig. 7.5(b). First presented are results for the 0.5 m resolution. Fig. 7.10 and Fig. 7.11 show the compressed point targets 1 and 9 accordingly. The resolution achieved in simulation for both presented point targets is 0.44 m in range direction and also 0.44 m in azimuth direction. Table 7.4 shows results for range and azimuth direction resolution for all nine point targets in the simulated scenario. The resolution of all nine point targets corresponds with the expected value. Images of both compressed point target 1 and 9 are symmetrical around the main lobe (see Fig. 7.10 and Fig. 7.11(a), (c) and (d)) and the phase characteristic of the point targets is linear (see Fig. 7.10 and Fig. 7.11(b), (e) and (f)). Both targets are correctly compressed with no visible distortions. The difference in resolution over all nine point targets is around 0.32% in range direction and 0.2% in azimuth direction normalized to the 'best' compressed point target. This is a very good result.
Results of a simulation with 0.3 m resolution for the point target 1 and 9 are presented in Fig. 7.12 and Fig. 7.13. The achieved resolution is 0.26 m for both targets in range and in azimuth direction. This corresponds with the expected value for this simulation. Images of both targets 1 and 9 are symmetrical in both range and azimuth direction. The phase characteristic is linear in both range frequency and azimuth frequency. This can be observed

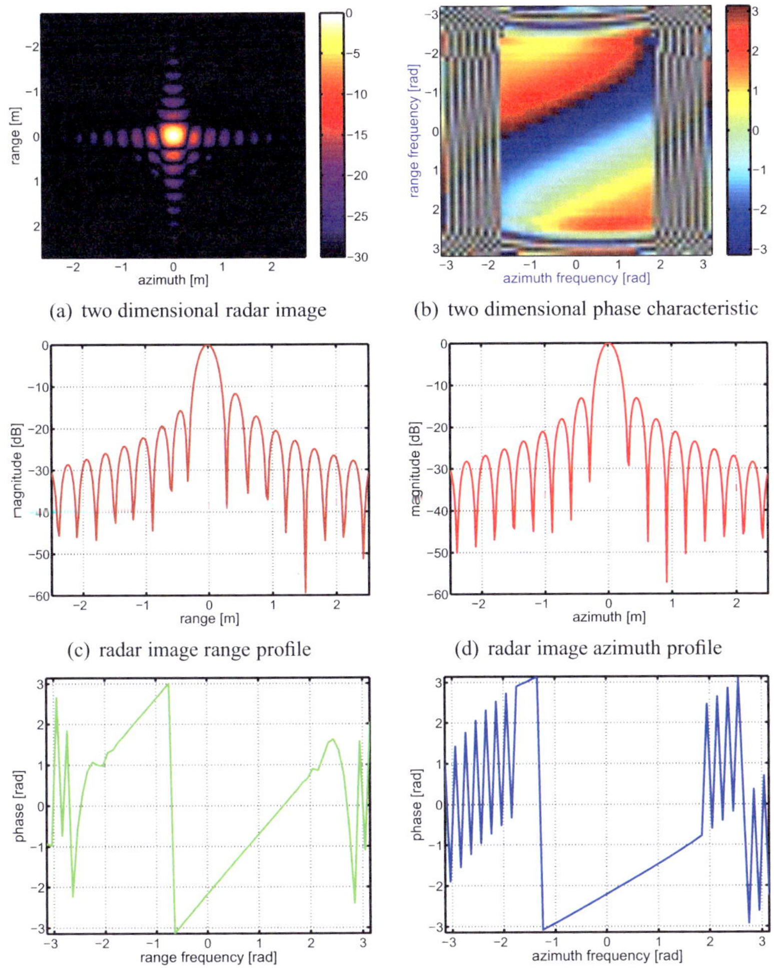

(a) two dimensional radar image

(b) two dimensional phase characteristic

(c) radar image range profile

(d) radar image azimuth profile

(e) phase characteristic cut in range frequency

(f) phase characteristic cut in azimuth frequency

Figure 7.8: Point target 1 with 0.3m resolution processed with the spaceborne chirp scaling algorithm.

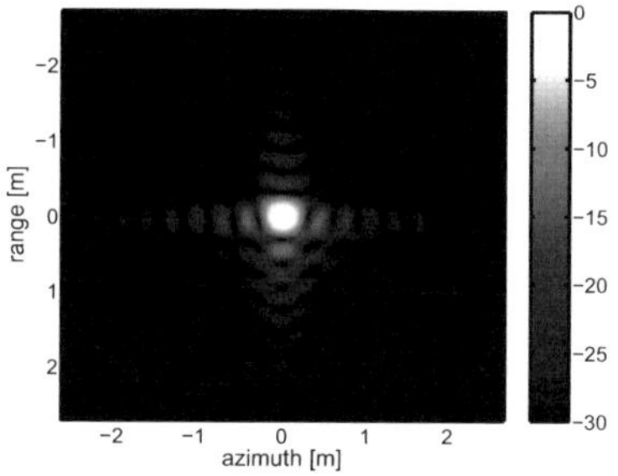

(a) two dimensional radar image

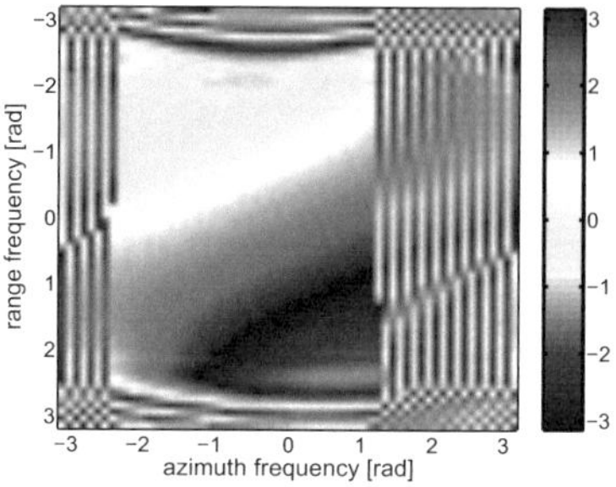

(b) two dimensional phase characteristic

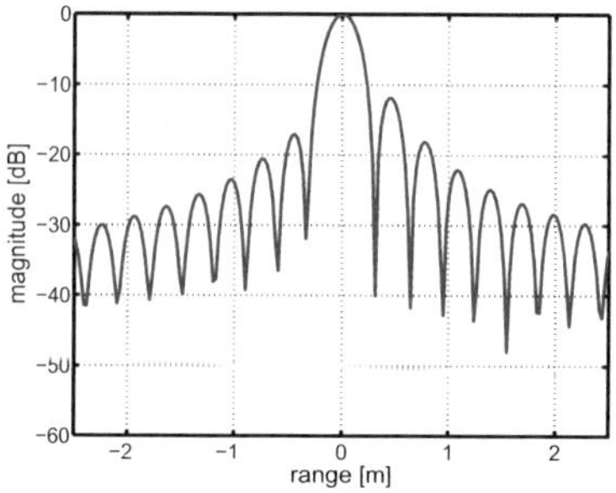

(c) radar image range profile

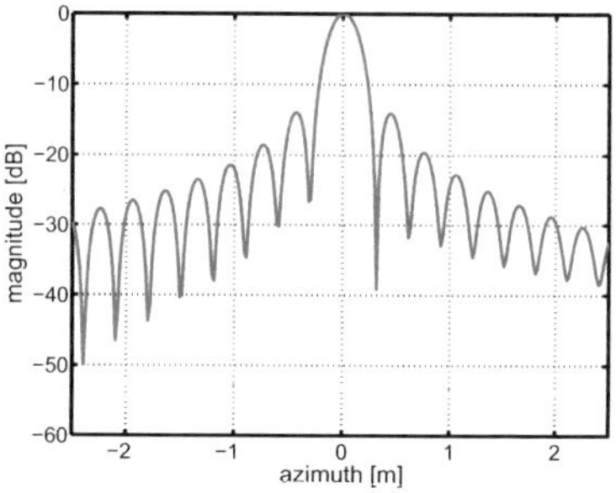

(d) radar image azimuth profile

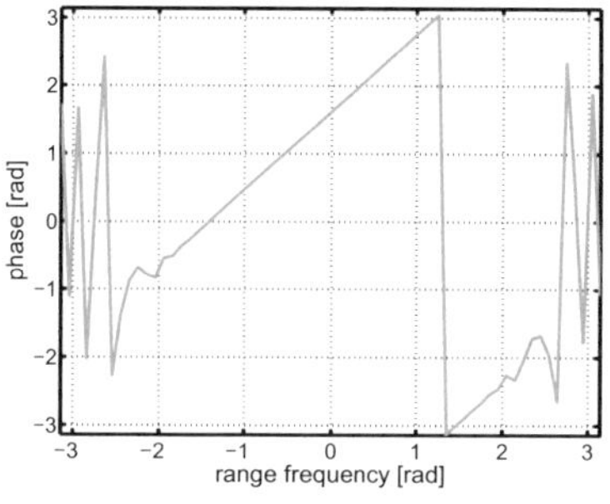

(e) phase characteristic cut in range frequency

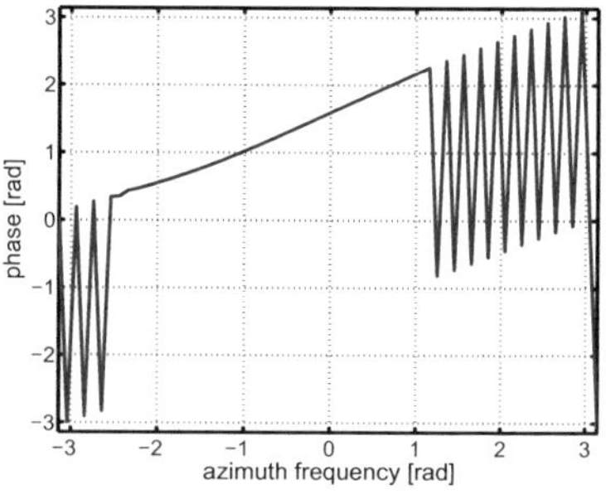

(f) phase characteristic cut in azimuth frequency

Figure 7.9: Point target 9 with 0.3m resolution processed with the spaceborne chirp scaling algorithm.

target no.	range pos. [km]	azimuth pos. [km]	range res. [m]	azimuth res. [m]
1	0	0	0.440	0.441
2	2.69	-2.69	0.440	0.441
3	2.69	0	0.439	0.441
4	2.69	2.69	0.439	0.441
5	0	-2.69	0.439	0.441
6	0	2.69	0.439	0.440
7	-2.69	-2.69	0.440	0.440
8	-2.69	0	0.440	0.440
9	-2.69	2.69	0.440	0.440

Table 7.4: Simulation for spaceborne omega-K algorithm for 0.5m resolution.

in the two dimensional phase characteristic as well as in the both cuts in range and azimuth frequency. Compared with results of the simulation with the same resolution compressed with the chirp scaling algorithm, the omega-K algorithm provides a better accuracy. Table 7.5 shows the resolution in range and in azimuth direction for all nine point targets in the simulated scene. The resolution or all nine targets corresponds nicely with the expected resolution for this simulation. The difference between the resolution between the 'best' and 'worst' point target is 0.46% for the range direction and 0.87% for the azimuth direction.

The omega-K algorithm provides more accuracy for the highly resolved SAR data compression. The radar image data compressed with the chirp scaling algorithm for 0.3 m resolution showed already first signs of performance loss due to inaccuracy of the processing algorithm. The same raw data compressed with the omega-K algorithm are fully focused with no signs of performance loss. A scenario with 0.2 m resolution was simulated to investigate if the omega-K algorithm can correctly compress such highly resolved radar data. The number of point targets in this scenario must had been limited to two due to very high amount of the simulation data and limitation in the processing power of the used hardware.

Results of a simulation with 0.2 m resolution in both range and azimuth direction compressed with the omega-K algorithm for both point targets 1a and 6a are presented in Fig. 7.14 and Fig. 7.15. The simulated scene consists only

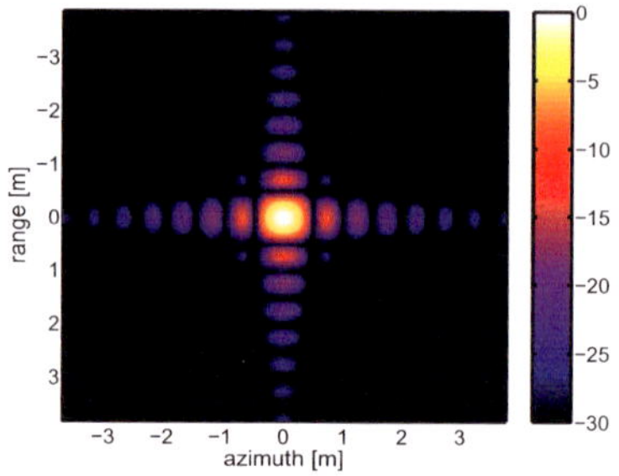

(a) two dimensional radar image

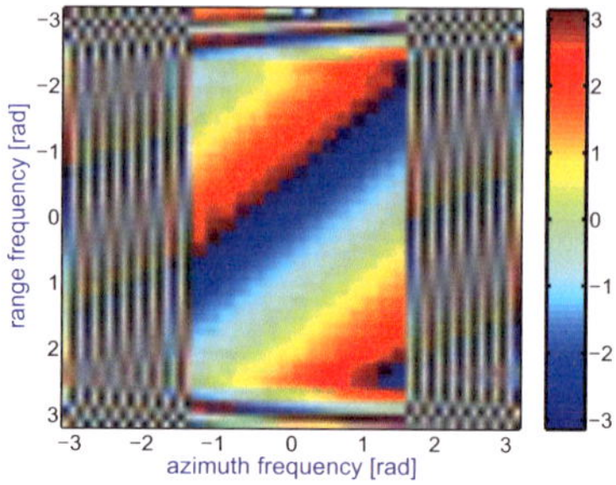

(b) two dimensional phase characteristic

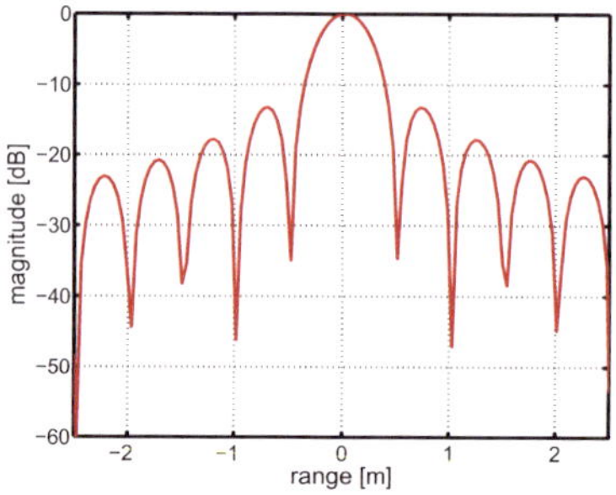

(c) radar image range profile

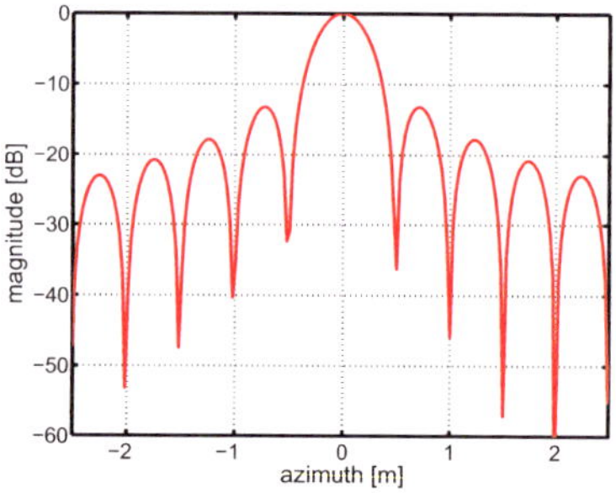

(d) radar image azimuth profile

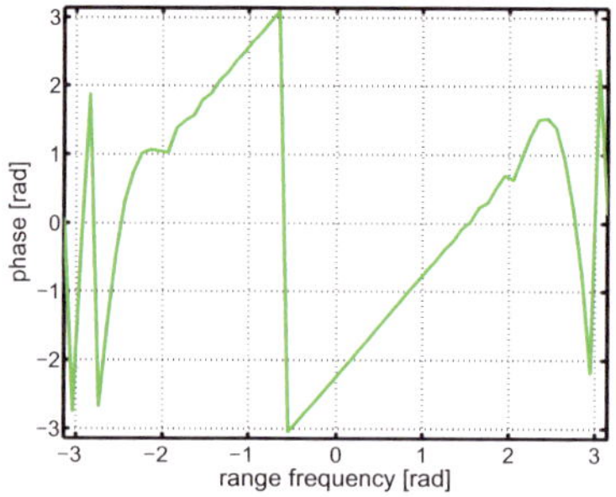

(e) phase characteristic cut in range frequency

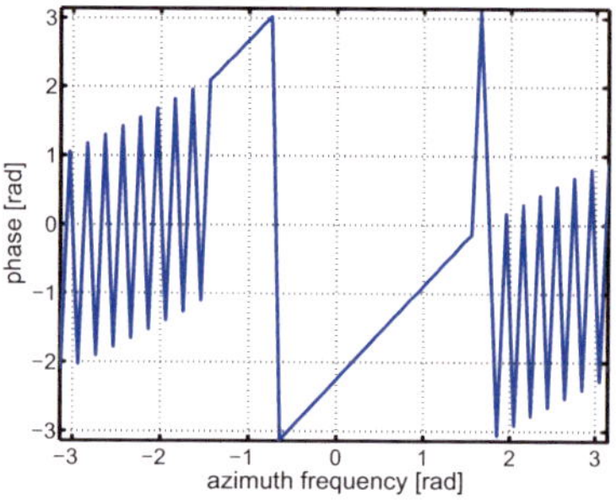

(f) phase characteristic cut in azimuth frequency

Figure 7.10: Point target 1 with 0.5 m resolution processed with the spaceborne omega-K algorithm.

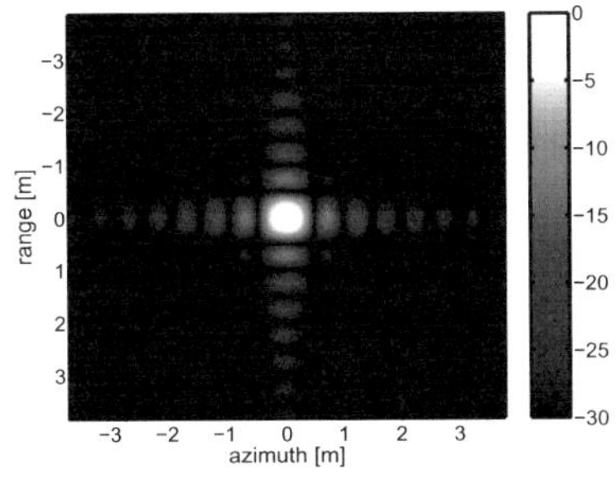

(a) two dimensional radar image

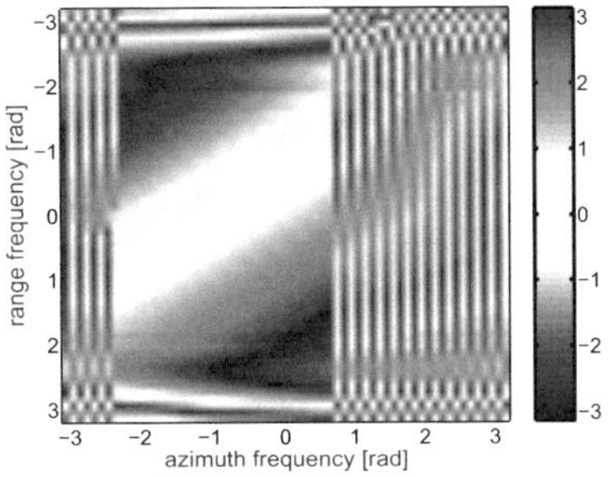

(b) two dimensional phase characteristic

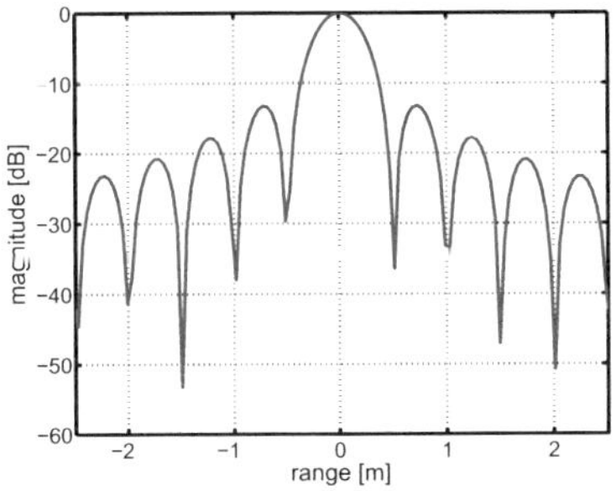

(c) radar image range profile

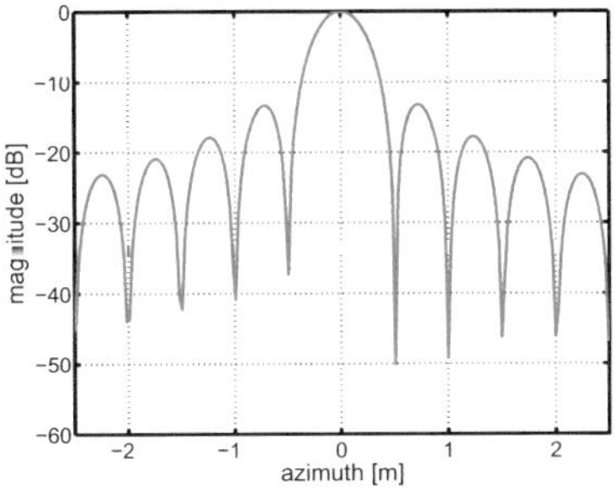

(d) radar image azimuth profile

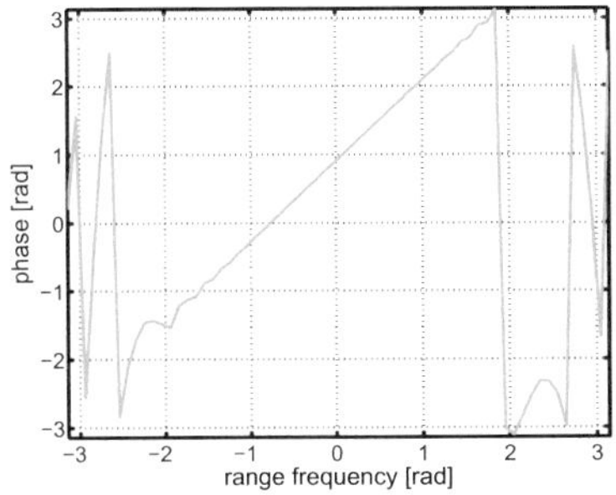

(e) phase characteristic cut in range frequency

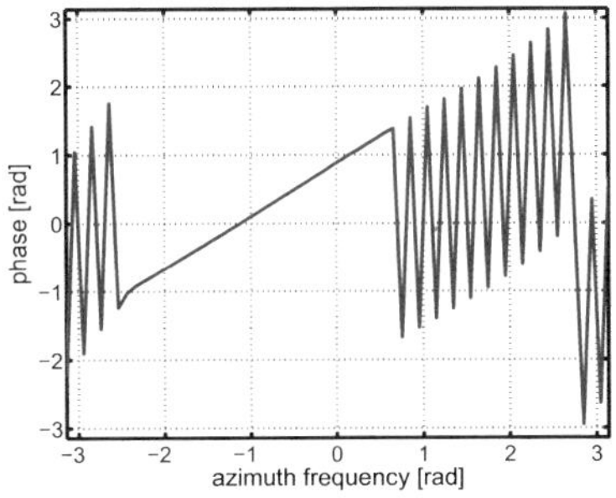

(f) phase characteristic cut in azimuth frequency

Figure 7.11: Point target 9 with 0.5 m resolution processed with the spaceborne omega-K algorithm.

target no.	range pos. [km]	azimuth pos. [km]	range res. [m]	azimuth res. [m]
1	0	0	0.264	0.265
2	2.69	-2.69	0.263	0.264
3	2.69	0	0.263	0.266
4	2.69	2.69	0.262	0.265
5	0	-2.69	0.263	0.264
6	0	2.69	0.263	0.265
7	-2.69	-2.69	0.263	0.264
8	-2.69	0	0.263	0.265
9	-2.69	2.69	0.263	0.265

Table 7.5: Simulation for spaceborne omega-K algorithm for 0.3m resolution.

of two point targets: $1a$ in the center of the scene and $6a$ in a position of about 2.6 km in azimuth from the center of the scene and 0 km in range direction. These points correspond in position to point targets 1 and 6 from the scene described in Section 6.7. The index a was added to indicate that only a part of the extended scenario has been simulated. The resolution achieved in the simulation for the target $1a$ is 0.22 m in range direction and 0.21 m in azimuth direction. For the point target $6a$ the resolution in range direction is 0.21 m and for the azimuth direction 0.22 m. Table 7.6 shows details for the range and azimuth resolution of simulated targets. The theoretically expected ideal resolution for the simulation is about 0.2 m with applied windowing which gives about 0.178 m resolution in both range and azimuth direction without window function applied. It can be easily calculated that the achieved resolution is about 21% worse than the theoretically expected value. The distortion can be clearly seen in Fig. 7.14 and Fig. 7.15. The point target in space domain is no longer symmetrical neither in range nor in azimuth direction. The two dimensional phase characteristic is not linear. Although there are strong distortions visible in the space domain and in the two dimensional phase characteristics, cuts in range frequency over the phase for both targets are linear. The cuts in the azimuth frequency over the phase characteristic show some slight curvature.

The omega-K algorithm in the form described in Section 7.2.1 can process spaceborne radar images with a resolution of about 0.3 m. It shows a better

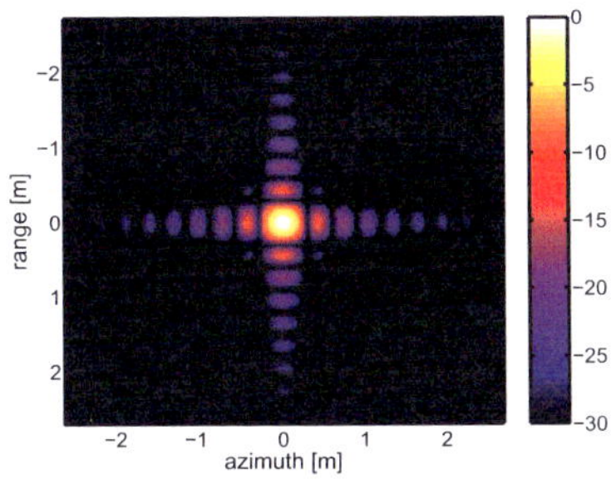

(a) two dimensional radar image

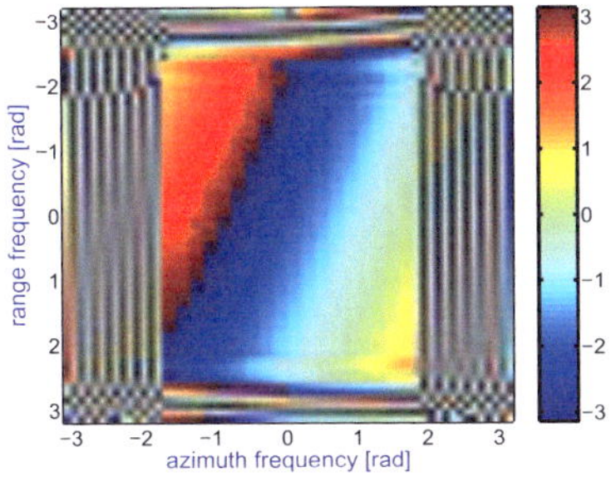

(b) two dimensional phase characteristic

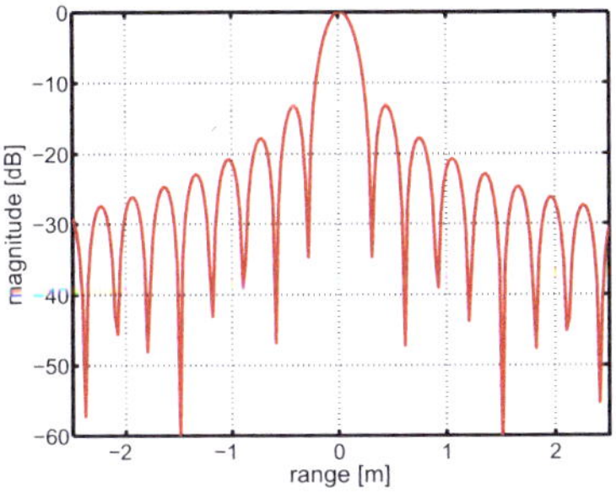

(c) radar image range profile

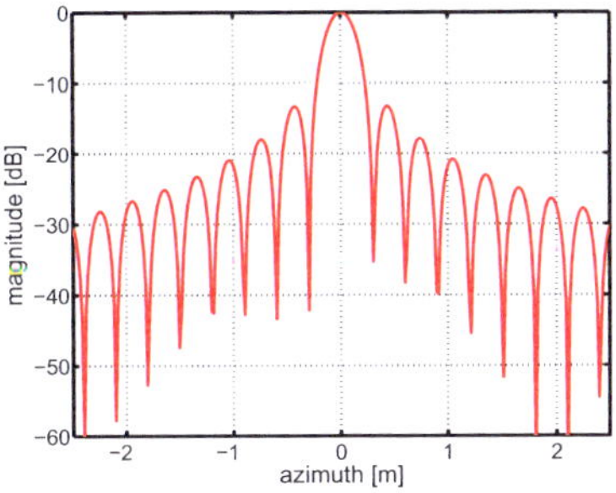

(d) radar image azimuth profile

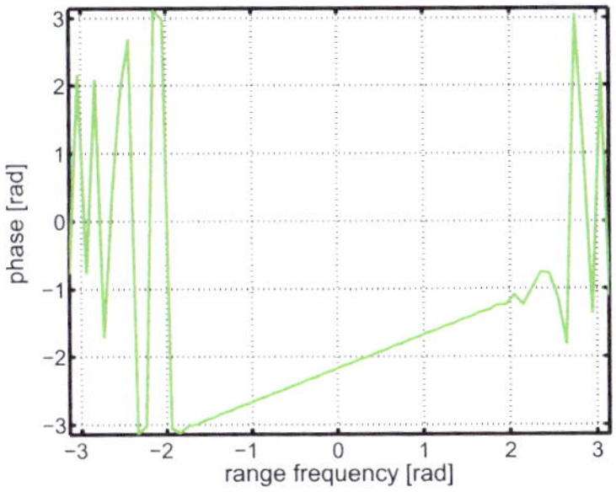

(e) phase characteristic cut in range frequency

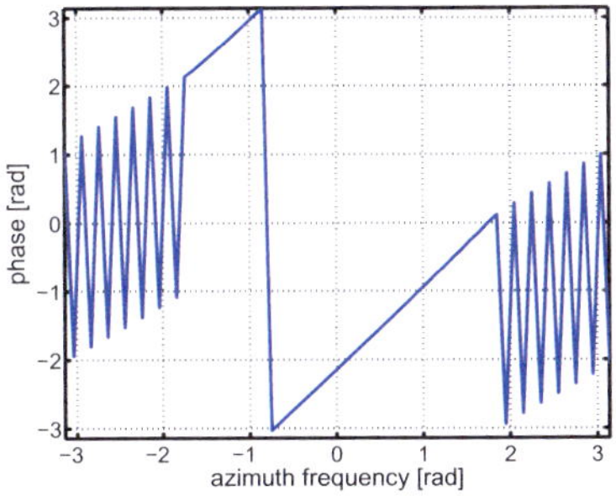

(f) phase characteristic cut in azimuth frequency

Figure 7.12: Point target 1 with 0.3 m resolution processed with the spaceborne omega-K algorithm.

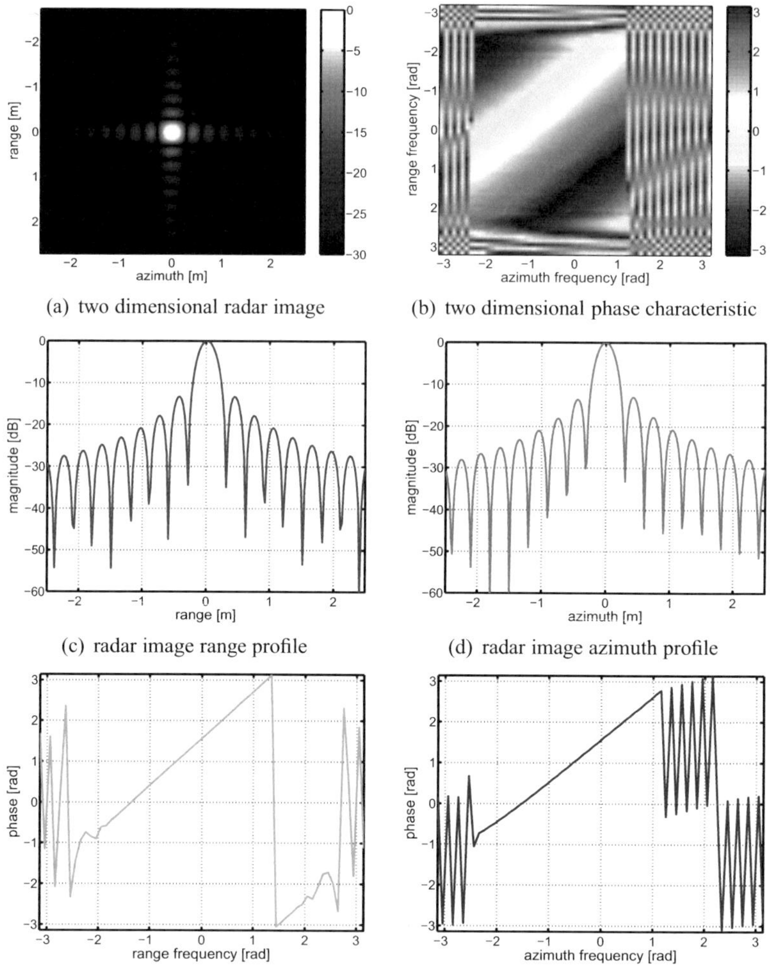

(a) two dimensional radar image

(b) two dimensional phase characteristic

(c) radar image range profile

(d) radar image azimuth profile

(e) phase characteristic cut in range frequency

(f) phase characteristic cut in azimuth frequency

Figure 7.13: Point target 9 with 0.3 m resolution processed with the spaceborne omega-K algorithm.

point target no.	1a	6a
range position [km]	0	0
azimuth position [km]	0	2.69
range resolution [m]	0.212	0.216
azimuth resolution [m]	0.215	0.214

Table 7.6: Simulation for spaceborne omega-K algorithm for 0.2m resolution.

performance for highly resolved radar data then the presented in previous paragraph chirp scaling algorithm. The algorithm is not accurate enough for processing SAR images with resolution of 0.2m and smaller. The degradation of about 20% in resolution meaning 0.22 m instead of 0.18 m might not seem to be severe. This is although not acceptable for a spaceborne SAR system.

7.4 Real Orbit Correction

The satellite trajectory of the radar platform described in the previous section of this chapter has been modeled as an ideal circular orbit. The Earth has been assumed to be a round sphere which is a very simplified model. This simple model is sufficient for a satellite SAR with moderate resolution. The inaccuracy of the processing algorithm begins to be noticeable in satellite SAR images with very high resolution and long illumination swath. A very precise estimation of the orbit needs to be considered in those cases. In reality the Earth is rather like a sphere flattened along the axis from the north pole to the south pole such that there is a bulge around the equator. This geometric figure is called oblate spheroid. The Earth has a radius of $R_{major} = 6378.137$ km along the major axis and a radius of $R_{minor} = 6356.7523$ km along the minor axis, [12]. This gives an aspect ratio, which is the ratio of the polar to the equatorial length, of 0.99664717 and a flattening of 0.003352859934. The Earth's surface is also not entirely flat. The highest point is the Mount Everest with 8848 meter above the local sea level. It is also possible that a part of land is underneath the local sea level like e.g. the area of the Dead Sea which is the worlds deepest depression with around 418 meters underneath the sea level. In case of a SAR satellite image with very high resolution the large changes in the height of the illuminated ground area must be also considered

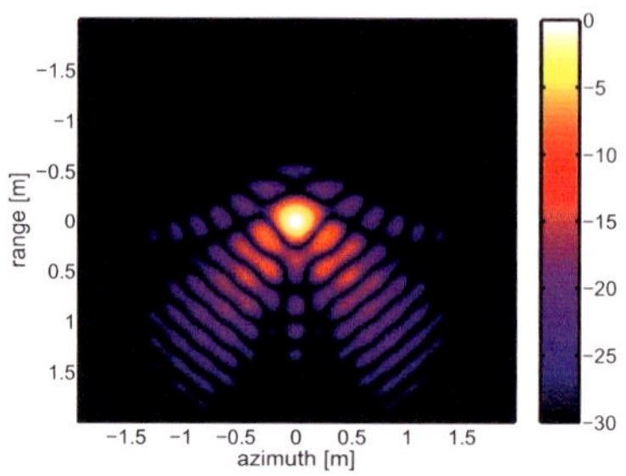

(a) two dimensional radar image

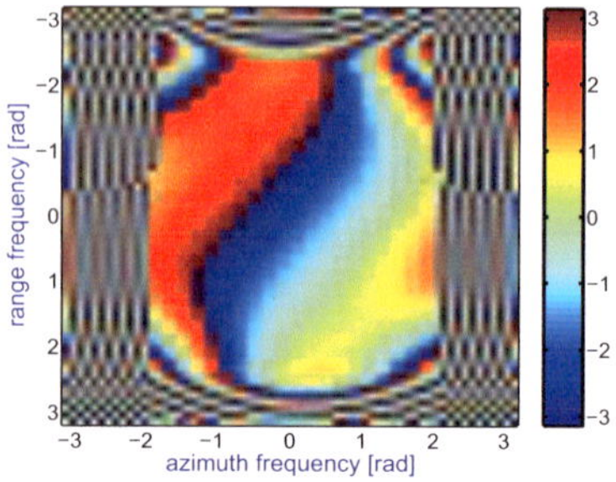

(b) two dimensional phase characteristic

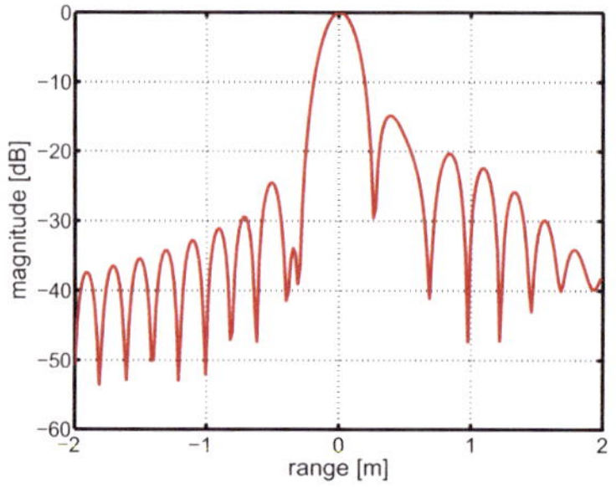

(c) radar image range profile

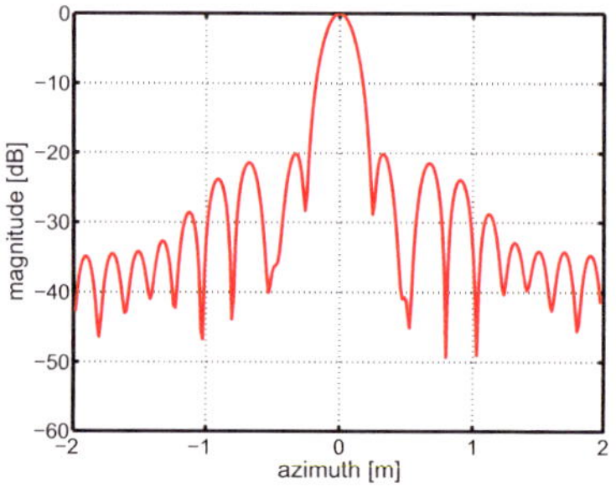

(d) radar image azimuth profile

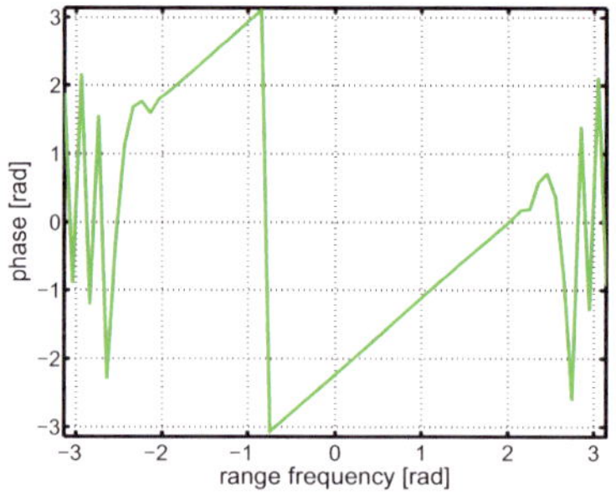

(e) phase characteristic cut in range frequency

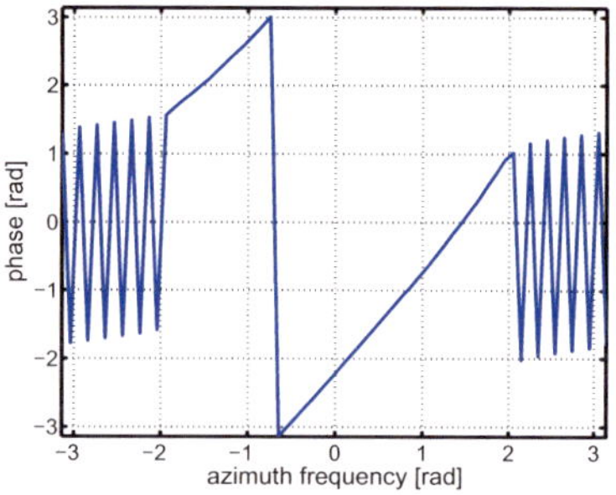

(f) phase characteristic cut in azimuth frequency

Figure 7.14: Point target $1a$ with 0.2 m resolution processed with the spaceborne omega-K algorithm.

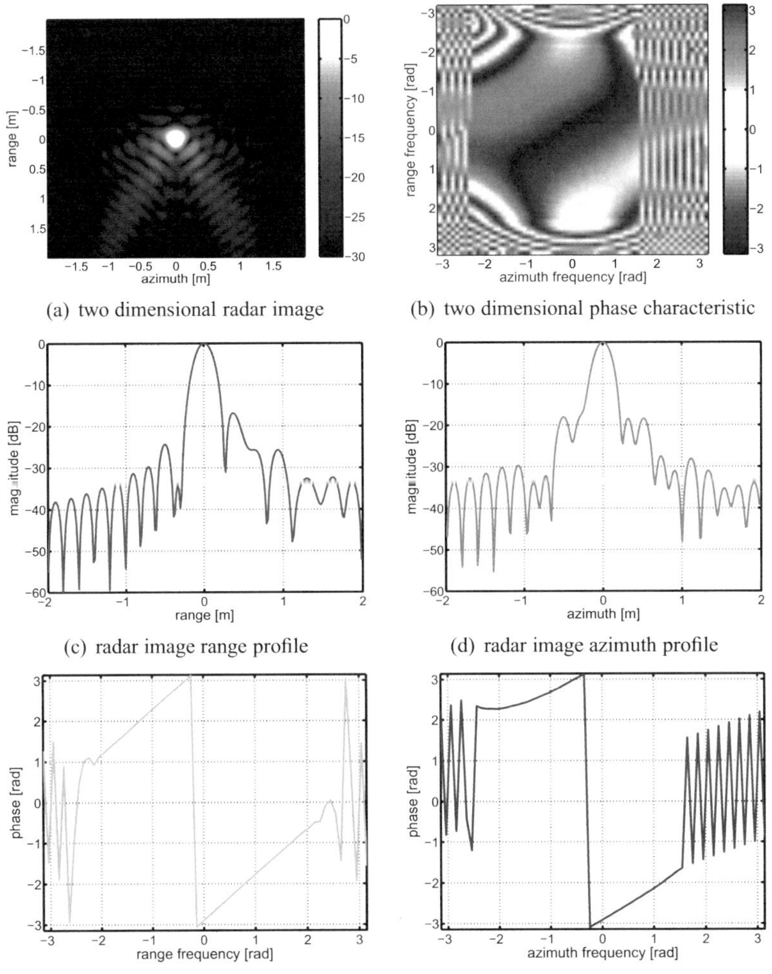

(a) two dimensional radar image

(b) two dimensional phase characteristic

(c) radar image range profile

(d) radar image azimuth profile

(e) phase characteristic cut in range frequency

(f) phase characteristic cut in azimuth frequency

Figure 7.15: Point target $6a$ with 0.2 m resolution processed with the spaceborne omega-K algorithm.

in the processing algorithm. The flattening of the Earth's surface influences also the shape of the satellite orbit. In first approximation the orbit can be assumed elliptic. The shape of the ellipse depends on the angle between the orbit plane and the plane going through the equator. The real orbit is much more complicated and takes account also of the irregularities in the Earth's surface, influence of the moon and more. The fluctuation of the satellite orbit can also have influence on a SAR image.

A correction of an arbitrary known satellite orbit implemented into the SAR processing is described in following. The difference between the ideal round orbit which is the basis for algorithms presented in this work and the real one can be seen as a position error of the radar platform. A processing for position error correction in the airborne case is presented first. On this base the position error correction for the spaceborne case algorithm is developed. The algorithm assumes the position error is known for all times of the data acquisition. Equations are adjusted to the spaceborne SAR system geometry in the following sections. The problem caused by a non-ideal trajectory of the radar platform is the defocussing of the resulting radar image. The SAR processing algorithm calculates a very large synthetic aperture by adding coherently receive signals from successive radar platform positions. Receive signals are added with phase shifts adequate to position of the platform at the acquisition time and assumed distance to the ground. Position errors change the phase of the receive signals resulting in phase errors in the generation of the synthetic aperture. This results in defocussing of the radar image as the added signals are no longer coherent.

7.4.1 Correction of the Radar Platform Trajectory for the Airborne Case

The ideal desired trajectory of an airborne SAR radar platform is a straight line which is the base of the SAR algorithms. The position of the plane can be measured during the flight with e.g. GPS system and be used for correction in the compression algorithm. If the position difference is known the phase difference between the ideal position and the real position can be compensated. The remaining from inaccuracy of the GPS measurement phase error can be compensated by an autofocus algorithms like e.g. mapdrift, phase gradient autofocus or space-variant refocus algorithm [8], [59].

An airborne SAR geometry, as presented in Fig. 7.16, is assumed for calculations. x_{ground} is the length on the ground between the projection of the radar platform and the center point of the illuminated swath. The reference distance equals to the mean slant range r_s with the incidence angle φ_i. Two separate cases with position error in z-direction or in x-direction are considered first, see in Fig. 7.16(a). In the first case the position error is defined by z_{me}. The resulting distance between the radar antenna and the point target in the center of the illuminated swath can be calculated from the property of a right-angle triangle.

$$r_s^2 = h^2 + x_{ground}^2 \tag{7.9}$$

Including the radar antenna with the position error in the z-direction of z_{me} the slant range distance with the position error $r_{me,z}$ can be calculated with Eq. (7.9).

$$r_{me,z}^2 = (h + z_{me})^2 + x_{ground}^2 \tag{7.10}$$

x_{ground} can be calculated from Eq. (7.9).

$$x_{ground}^2 = r_s^2 - h^2 \tag{7.11}$$

By applying Eq. (7.11) into Eq. (7.10):

$$r_{me,z} = \sqrt{(h + z_{me})^2 + r_s^2 - h^2} \tag{7.12}$$

The error in the distance between the point target and the radar antenna $\Delta r_{me,z}$ can be calculated as the difference between Eq. (7.9) and Eq. (7.12).

$$\Delta r_{me,z} = r_s - r_{me,z} = R_s - \sqrt{(h + z_{me})^2 + r_s^2 - h^2} \tag{7.13}$$

A similar approach can be applied for a position error in the x-direction of x_{me}, see Fig. 7.16(a). The distance between the radar antenna and the center of the illuminated swath with a position error of the antenna in x-direction $r_{me,x}$ is to be calculated according to Eq. 7.14.

$$\begin{aligned} r_{me,x} &= \sqrt{h^2 + (x_{ground} - x_{me})^2} \\ &= \sqrt{h^2 + \left(\sqrt{r_s^2 - h^2} - x_{me}\right)^2} \end{aligned} \tag{7.14}$$

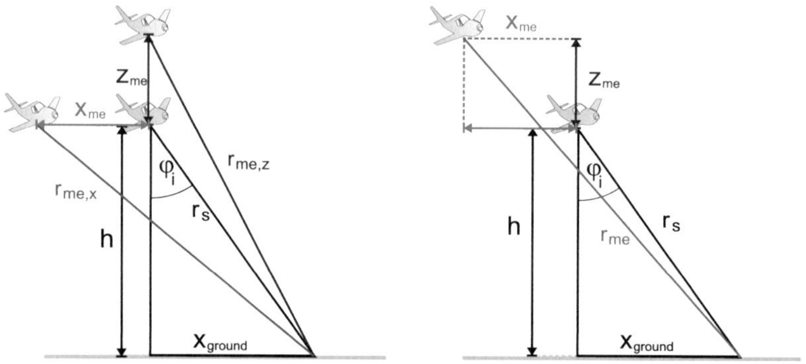

(a) position error in x- or z-direction (b) position error in both x- and z-direction

Figure 7.16: Geometry of the airborne SAR with position errors.

The difference between distances without and with the position error $\Delta r_{me,x}$ can be calculated from Eq. (7.9) and Eq. (7.14).

$$\Delta r_{me,x} = r_s - r_{me,x} = r_s - \sqrt{h^2 + \left(\sqrt{r_s^2 - h^2} - x_{me}\right)^2} \tag{7.15}$$

Combining Eq. (7.12) and Eq. (7.14) for position errors in both x- and z-direction results in Eq. 7.17.

$$r_{me} = \sqrt{\left(\sqrt{r_s^2 - h^2} - x_{me}\right)^2 + (h + z_{me})^2} \tag{7.16}$$

and the difference between the ideal trajectory and the real trajectory for this position Δr_{me} is equal to:

$$\Delta r_{me} = r_s - \sqrt{\left(\sqrt{r_s^2 - h^2} - x_{me}\right)^2 + (h + z_{me})^2} \tag{7.17}$$

This difference can be easily converted to a phase difference for the specified point target between the ideal trajectory and the real trajectory of the radar platform.

7.4.2 Position Errors in the Spaceborne Case

A cut in elevation of a spaceborne SAR system geometry is shown in Fiq. 7.17. The position error in x-direction x_{me} and in z-direction z_{me} are marked in this figure. h is the altitude of the ideal orbit. x_{mc} is the base of a right-angle triangle between the radar position and the middle of the illuminated swath on the ground and h_{mc} is the height of this right-angle triangle as shown in Fig. 7.17. It is to be noted that with the definition of the right-angle triangle between the radar antenna and the illuminated ground point an equation similar to the airborne case can be used for the calculation of the spaceborne case. x_{mc} can be calculated from the equation for the area of the triangle. Eq. (7.18) shows the Heron equation for triangles area.

$$P = \frac{1}{4}\sqrt{(a+b+c)(a+b-c)(a-b+c)(-a+b+c)} \tag{7.18}$$

Where a, b and c are the side lengths of the triangle. Using Eq. (7.18) x_{mc} can be calculated as:

$$\begin{aligned} x_{mc} &= \frac{\sqrt{(2R_{Earth}+h+r_s)(h+R_{Earth})(r_s-h)}}{2(h+R_{Earth})}\cdot\ldots \\ &\quad \sqrt{(2R_{Earth}+h-r_s)} \\ &= \frac{1}{2}\frac{\sqrt{((2R_{Earth}+h)^2-r_s^2)\left(r_s^2-h^2\right)}}{h+R_{Earth}} \end{aligned} \tag{7.19}$$

h_{mc} can be calculated as:

$$h_{mc} = \sqrt{r_s^2 - x_{mc}^2} = (R_{Earth}+h) - \sqrt{R_{Earth}^2 - x_{mc}^2} \tag{7.20}$$

The resulting distance r_{me} between the point target and the radar antenna with position errors in x- and z-direction can be calculated according to Eq. (7.21).

$$r_{me} = \sqrt{\left(\sqrt{r_s^2 - h_{mc}^2} - x_{me}\right)^2 + (h_{mc}+z_{me})^2} \tag{7.21}$$

The difference Δr_{me} between the reference slant range and the reference slant range with position errors of the satellite radar antenna can be calculated according to Eq. (7.22). A correction phase and range shift can be calculated based on the difference of the slant range. The correction is then applied in

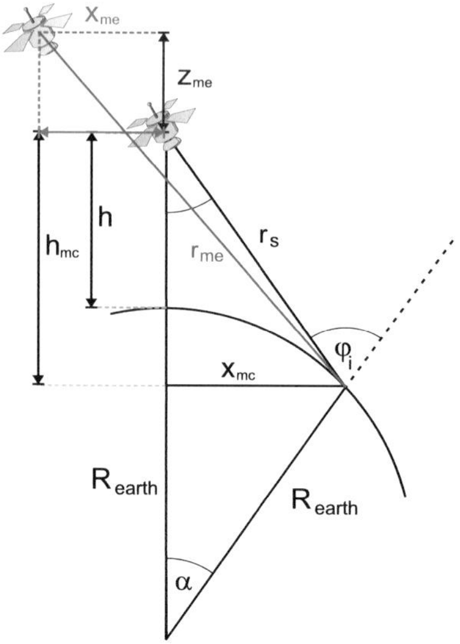

Figure 7.17: Geometry of the satellite SAR with position errors.

the SAR image processing. A detailed description of the correction in the chirp scaling and the omega-K algorithms follows in next section.

$$\Delta r_{me} = r_s - \sqrt{\left(\sqrt{r_s^2 - h_{mc}^2} - x_{me}\right)^2 + (h_{mc} + z_{me})^2} \tag{7.22}$$

The position error influence was calculated up to now only for the reference range. In order to correctly focus the whole SAR image Δr_{me} has to be calculated for all ranges. Through replacing the reference range r_s with the range vector r (see Eq. (7.25)) the resulting error in distance can be calculated for every range cell of the image. It can be easily proven that this equation is valid also for other range values. Eq. (7.25) describes the difference in the slant rage calculated for all ranges of the illuminated area.

$$x_{mc}(r) = \frac{1}{2}\frac{\sqrt{((2R_{Earth} + h)^2 - r^2)(r^2 - h^2)}}{h + R_{Earth}} \tag{7.23}$$

$$h_{me}(r) = (R_{Earth} + h) - \sqrt{R_{earth}^2 - x_{mc}(r)^2} \tag{7.24}$$

$$\Delta r_{me}(r) = r - \sqrt{\left(\sqrt{r^2 - h_{mc}(r)^2} - x_{me}\right)^2 + (h_{mc}(r) + z_{me})^2} \quad (7.25)$$

$$\Delta r_{me}(r) = r - \sqrt{(x_{mc}(r) - x_{me})^2 + (h_{mc}(r) + z_{me})^2} \quad (7.26)$$

The range r can be easily recalculated into a fast time t_{rg}.
In high resolution radar images the height of the terrain should also be considered in processing. The difference in height may lead to defocussing in the radar image [37]. h_{ground} is the height of the illuminated terrain. Using the same principle that was used to calculate the influence of the satellite position error the following equations are achieved.

$$x_{mc}(r) = \frac{1}{2}\frac{\sqrt{((2R_{Earth} + h + h_{ground})^2 - r^2)(r^2 - h^2)}}{h + R_{Earth} + h_{ground}} \quad (7.27)$$

$$\begin{aligned} h_{mc}(r) &= (R_{Earth} + h + h_{ground}) - \ldots \\ &\sqrt{(R_{Earth} + h_{ground})^2 - x_{mc}(r)^2} \end{aligned} \quad (7.28)$$

$$\Delta r_{me}(r) = r - \sqrt{\left(\sqrt{r^2 - h_{mc}(r)^2} - x_{mc}(r)\right)^2 + (h + z_{mc})^2} \quad (7.29)$$

$$\Delta r_{me}(r) = r - \sqrt{(x_{mc}(r) - x_{mc}(r))^2 + (h + z_{mc})^2} \quad (7.30)$$

7.4.3 Implementation of the Position Error Correction

The implementation of a correction for position errors of the radar platform in the spaceborne SAR processing algorithm is described in the following. The calculation of the difference in distance between the ideal radar platform trajectory and a trajectory with position error in x- and z-axis was presented in the previous section. So far a position error has been considered for a single position in azimuth direction. The SAR processing of a single ground point is based on the processing of receive radar signals from a very long synthetic aperture. The position error of all considered positions of the radar antenna has influence on the processing of a single ground point and needs to be corrected in the radar processing. It is to be noted that position errors vary for every processed ground point as a slightly different synthetic aperture is constructed during processing for each radar image point.

The proposed algorithm with the position error correction consists of two steps. First the position error is corrected in the frequency domain. This first correction stage is range independent as the error is estimated for a single specified range usually in the center of the illuminated swath. The second correction in the time domain corrects the motion errors for all range bins.
The first correction is applied on the range compressed data. The data is transformed into the frequency-time domain and a multiplication with an adequate range reference function is performed. The data can then be multiplied with the motion error correction factor in the frequency domain. The correction is calculated according to Eq. (7.31).

$$\Phi_{mc,f}(r_{ref}, u) = \exp\left\{j\frac{2\pi}{\Delta r}\Delta r_{me}(r_{ref}, u)\right\} \tag{7.31}$$

The equation for the motion error depends on the azimuth position u. The correction is calculated for the reference range r_{ref} preferably in the center of the illuminated swath. This multiplication shifts the range lines of the radar data in the correct position.
The second correction for the motion error of the radar platform takes place in the time domain. The radar data is transformed back into the time domain with the inverse Fourier transformation in range direction. The range compressed radar data is multiplied with the position correction function calculated according to Eq. (7.32) which precisely corrects the motion error for every range cell.

$$\Phi_{mc,t}(r, u) = \exp\left\{j\frac{4\pi}{\lambda}\Delta r_{me}(r, u)\right\} \tag{7.32}$$

The second correction function depends on azimuth u and on range r. The phase error caused by the position error is corrected for every image point with this multiplication. After this position correction the radar data is processed further with one of the SAR processors presented in the previous sections. As the radar data was already range compressed at the beginning of the processing chain functions in the further processing must be modified accordingly.
The modification in the chirp scaling algorithm can be applied in the function described with Eq. (6.29) which introduces the correction of the bulk RCMC and range compression. The modified function can be seen in Eq. (7.33).

This function corrects the bulk RCMC but provides no range focus. Fig. 7.18 shows the flow chart for the modified chirp scaling algorithm.

$$\Phi_2(f_{rg}, f_{az}, r_{ref}) = \exp\left(-j\pi\frac{f_{rg}^2}{K_s(f_{az}, r_{ref})(1 + C_s(f_{az}))} + \dots \right.$$
$$\left. j\frac{4\pi}{c_0}f_{rg}r_{ref}C_s(f_{az}) + j\frac{\pi f_{rg}^2}{k_r}\right) \quad (7.33)$$

The change for the omega-K algorithm concerns the reference function described by Eq. (6.34). A similar modification, where the range compression part of the equation is removed, is presented in Eq. (7.34).

$$\theta(f_{rg}, f_{az}) = \frac{4\pi R_{ref}}{c_0}\sqrt{(f_{rg} + f_c)^2 - \frac{c_0^2 f_{az}^2}{4v_s^2}} \quad (7.34)$$

In the sliding spotlight mode the motion error correction takes place before the azimuth filter. The implementation of the correction after the azimuth filter would be much more complicated as the data is recalculated to an equivalent strip-map mode geometry. The flow chart of the omega-K algorithm with implemented correction of position errors is presented in Fig. 7.19.

7.5 Simulation Results for Real Orbit Correction

The omega-K algorithm with position error correction have been verified with simulations. The scenario presented in Section 7.3 was used to create a raw data for sliding spotlight mode including the position error of the satellite flight trajectory. The simulated resolution was 0.5 m in both range and azimuth direction. Parameters for corresponding radar system can be found in Table 7.1. The simulated raw radar data was processed once with the omega-K algorithm for sliding spotlight mode without position error compensation and one with the additional correction. The position error was applied in height, which corresponds to the z coordinate, with a maximum error of ± 0.5 m. It has been assumed that the position error is known for all radar positions.

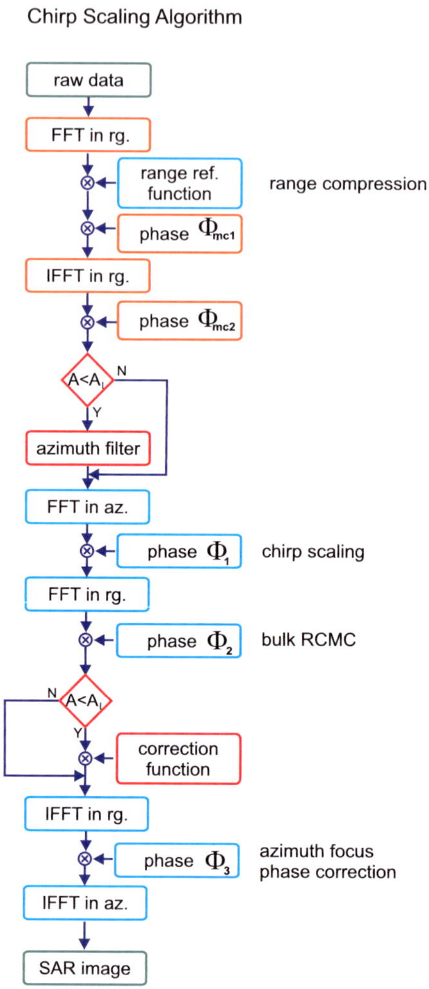

Figure 7.18: Flow chart of chirp scaling algorithm spaceborne processor for sliding spotlight mode

Fig. 7.20 shows results of the simulation for the compression algorithm without any correction for motion errors. The point target 1 in the center of the

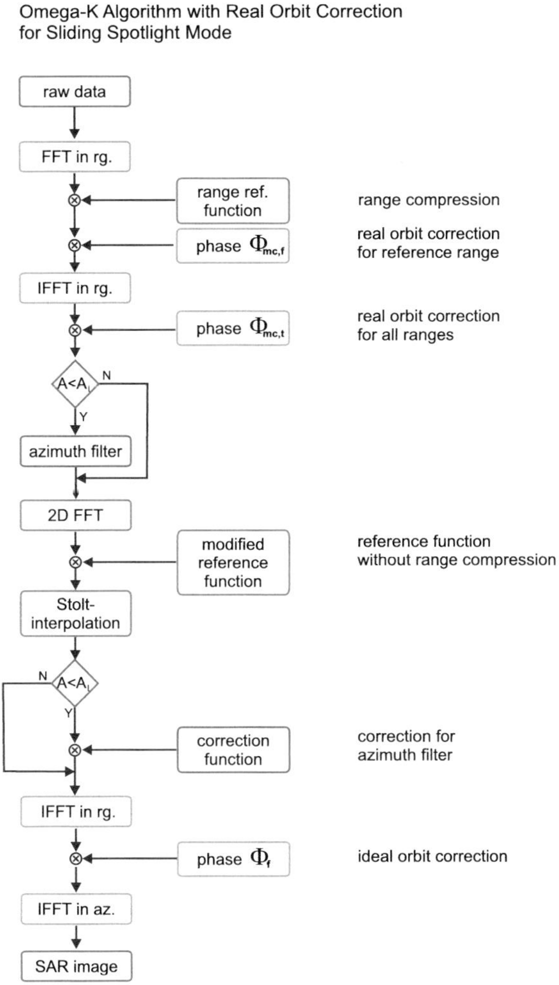

Figure 7.19: Flow chart of omega-K algorithm spaceborne processor for sliding spotlight mode

swath in range and azimuth is presented in this figure. Figure (a) shows the magnitude in the space domain, (b) the two-dimensional phase characteristic of the point target, (c) shows the cuts in range of the magnitude in space domain and (d) in azimuth direction, (e) shows the cut in range frequency and (e) in azimuth frequency of the phase characteristic. The point target is strongly defocused in azimuth direction which can be seen in Fig. 7.20. The point target response is spread over about 60 meters. Because of the very wide spread of the compressed point target the resolution was not analyzed for this simulation. The phase of the spectrum is clearly not linear in azimuth frequency. The remaining eight point targets are also defocused in similar way.

Results for the compression algorithm with position error correction for the point target 1 positioned in the center of the swath are presented in Fig. 7.21. In this simulation the same raw data was processed with the omega-K algorithm with position error correction. The calculated resolution for the point target 1 is 0.44 m in range and 0.44 m in azimuth. This corresponds with the results achieved for the omega-K algorithm for a simulation without position errors for the same resolution. The point target is correctly compressed. The point target in space domain is symmetrical in both range and azimuth direction. The phase characteristics of this target is linear. The processing algorithm with the compensation of non-ideal orbit is capable of focusing the raw data. No degradation of system performance is present. It is possible that although the center point target 1 is fully compressed the focus is lost for targets positioned farther away from the center of the scene. The point target 9 is positioned about -2.6 km from the center of the swath in range and about 2.69 km in azimuth direction. Fig. 7.22 shows the results for the point target 9. Plots in Fig. 7.22 show that the position error correction algorithm works also for ranges different from the reference range r_{ref} and also for azimuth positions different from the swath center. The point target in space domain is symmetrical in both range and azimuth direction. The phase characteristic is linear in range frequency and in azimuth frequency. The resolution for the point target 9 is 0.44 m in range and 0.44 m in azimuth direction. Table 7.7 shows results for range and azimuth resolution for all nine point targets positioned in the simulated scenario. The resolution is consistent with the expected value calculated from the system parameters for all nine targets.

It was shown that the position error correction algorithm works very good with the spaceborne sliding spotlight algorithm. Comparing the results for a simulation for 0.5 m resolution compressed with the omega-K algorithm

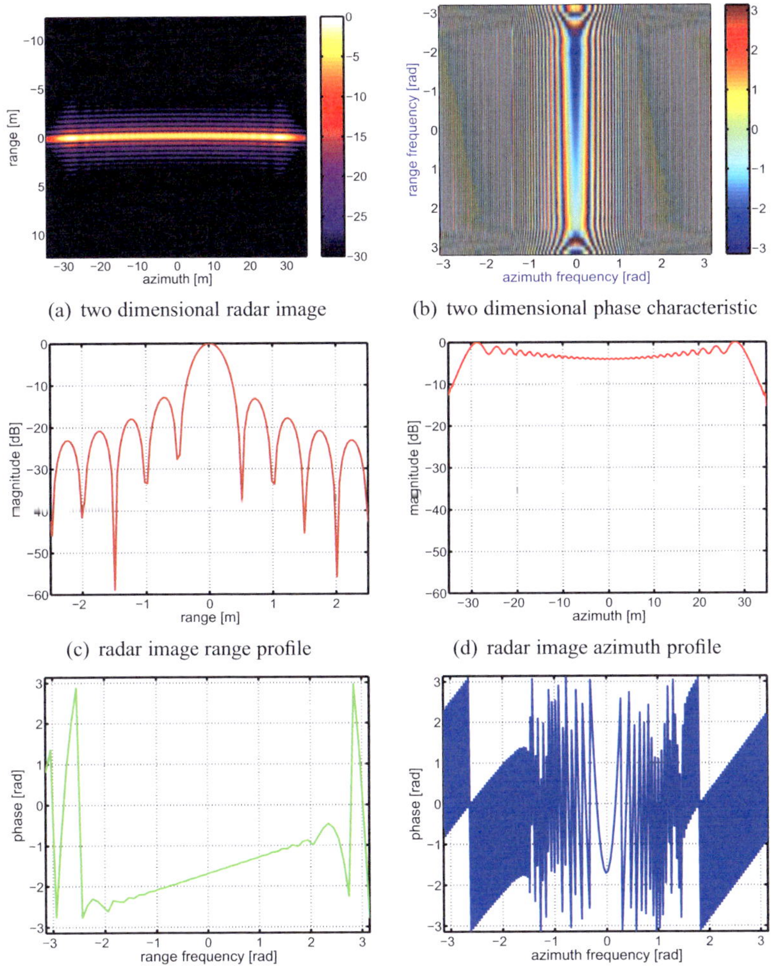

(a) two dimensional radar image

(b) two dimensional phase characteristic

(c) radar image range profile

(d) radar image azimuth profile

(e) phase characteristic cut in range frequency

(f) phase characteristic cut in azimuth frequency

Figure 7.20: Point target 1 with 0.5 m resolution with motion errors compressed with the spaceborne omega-K algorithm.

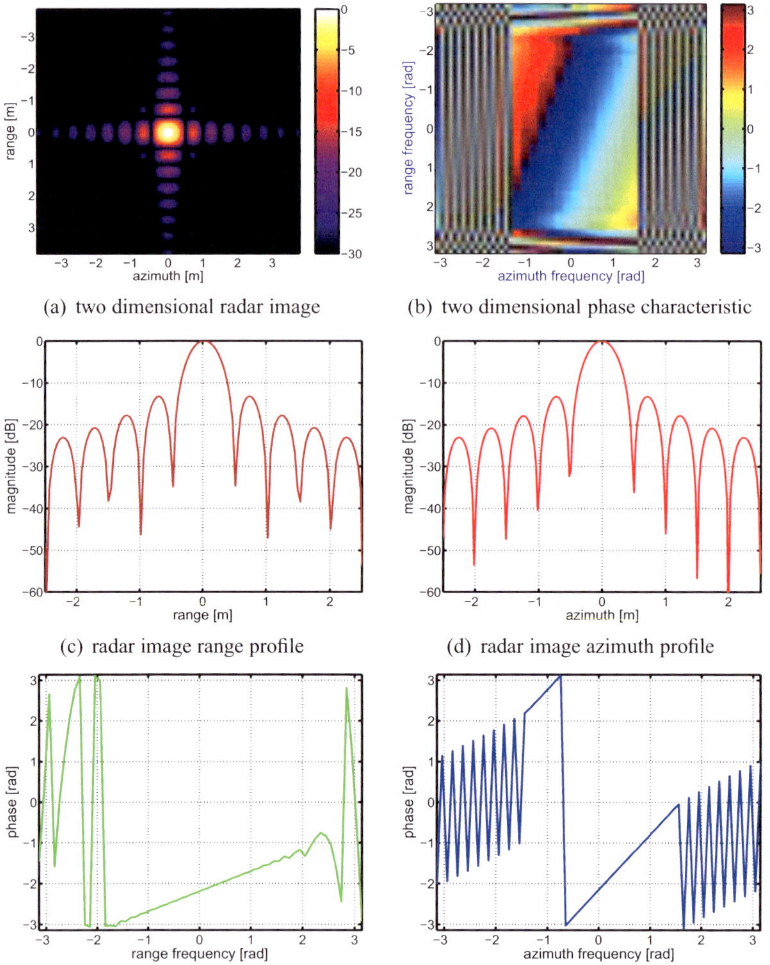

(a) two dimensional radar image

(b) two dimensional phase characteristic

(c) radar image range profile

(d) radar image azimuth profile

(e) phase characteristic cut in range frequency

(f) phase characteristic cut in azimuth frequency

Figure 7.21: Point target 1 with 0.5 m resolution compressed with the spaceborne omega-K algorithm with motion errors correction.

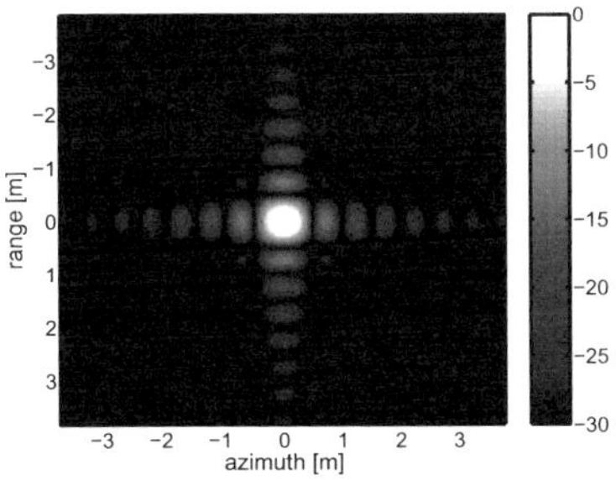

(a) two dimensional radar image

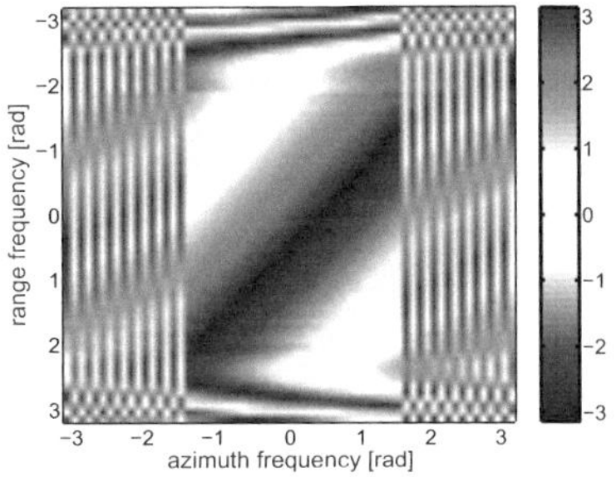

(b) two dimensional phase characteristic

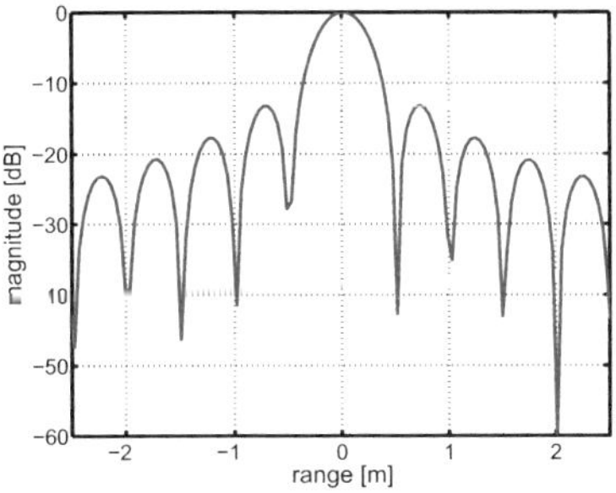

(c) radar image range profile

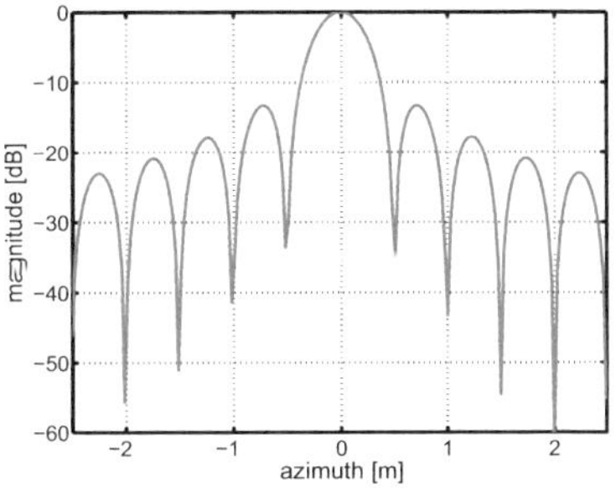

(d) radar image azimuth profile

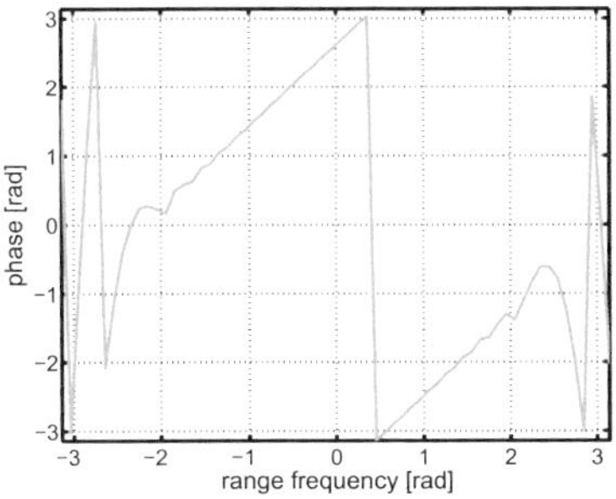

(e) phase characteristic cut in range frequency

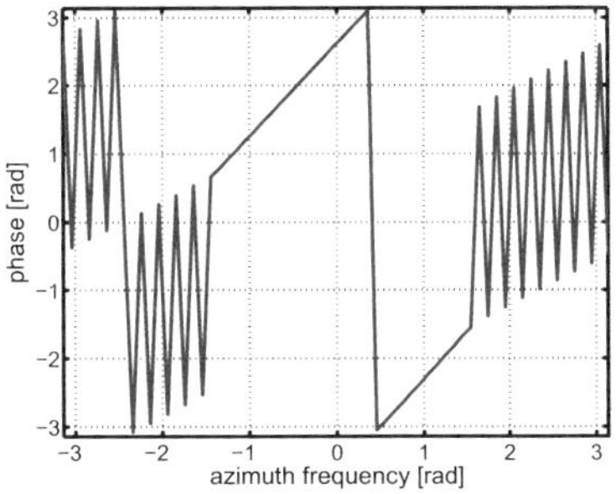

(f) phase characteristic cut in azimuth frequency

Figure 7.22: Point target 9 with 0.5 m resolution compressed with the spaceborne omega-K algorithm with motion errors correction.

target no.	range pos. [km]	azimuth pos. [km]	range res. [m]	azimuth res. [m]
1	0	0	0.440	0.440
2	2.69	-2.69	0.440	0.441
3	2.69	0	0.439	0.441
4	2.69	2.69	0.439	0.441
5	0	-2.69	0.441	0.441
6	0	2.69	0.439	0.440
7	-2.69	-2.69	0.440	0.440
8	-2.69	0	0.440	0.440
9	-2.69	2.69	0.440	0.440

Table 7.7: Simulation for the spaceborne omega-K algorithm with motion correction for 0.5 m resolution.

presented in Section 7.3 and the simulation including position errors compressed with the omega-K algorithm including position error compensation no decrease of performance can be seen.

7.6 Summary

The processing of the spaceborne sliding spotlight SAR data was the concern of this chapter. Two algorithms for processing of the airborne sliding spotlight SAR data: the chirp scaling algorithm and the omega-K algorithm were presented and evaluated in the previous Chapter 6. In this chapter both algorithms were adapted to process the spaceborne SAR data. The performance of proposed processing algorithms has been evaluated by the means of simulation. The differences in geometry of data acquisition between the airborne and the spaceborne sliding spotlight mode were investigated. A modification for the sliding spotlight mode parameter adapting it to a spaceborne case has been proposed. The modified parameter maintained the properties of the parameter defined for the airborne sliding spotlight mode case. The simple spaceborne trajectory of an ideal circular Earth and orbit were assumed at first for modeling raw radar data. The velocity ratio parameter necessary to correctly compress the spaceborne SAR raw data was calculated. The chirp scaling and the omega-K algorithms were modified to accommodate for the spaceborne sliding spotlight SAR mode. The performance of proposed algorithms was verified in simulation with various resolutions. The spaceborne chirp scaling algorithm can process the high resolution SAR sliding spotlight mode data. This algorithm is very efficient in processing power as it consists only of FFTs and multiplications. The chirp scaling algorithm processed the simulated SAR raw data with 0.5 m and with 0.3 m resolution. The simulation with 0.3 m showed some minor distortion due to inaccuracy of the compression processing. The chirp scaling algorithm is still applicable for radar data with this high resolution. For higher resolution a more accurate processing algorithm should be used. The same scenario simulated with 0.5 m and with 0.3 m resolution has been used to investigate the performance of the omega-K algorithm. In this case the compressed with the omega-K algorithm raw data with 0.3 m resolution showed no decrease of performance. The spaceborne omega-K algorithm is more applicable for processing highly resolved SAR data. The accuracy of the algorithm is higher compared to the chirp scaling algorithm. A raw data with a resolution of 0.2 m has been generated to evaluate the performance of the omega-K algorithm for those high resolutions. In the compressed SAR data some distortions due to inaccuracies of the processing were visible. The achieved performance is not acceptable for a spaceborne SAR system.

The assumption of a round orbit is an ideal case. A proposal for correction of real arbitrary orbit was described. The difference between the ideal round orbit and the real trajectory flown by a satellite can be treated as a position error of the radar platform. A correction algorithm proposal for this position error allowing an arbitrary trajectory of the radar platform was described. The chirp scaling algorithm and the omega-K algorithm were modified accordingly to compensate for the non-ideal flight trajectory. The flow charts for both algorithms were presented. The performance of the modified omega-K algorithm was investigated with the help of simulation. The test raw data was correctly compressed with no visible deformation or defocus in the image radar data. The omega-K algorithm with position correction is adequate for the highly resolved spaceborne sliding spotlight SAR data.

8 Conclusion

The general trend in development of imaging radar is to provide higher resolution, larger coverage area and shorter revisit time. More flexibility of radar systems is demanded for various applications. The requirements for civil SAR application are mainly large coverage area with moderate resolution. The same area is often revisited multiple times to monitor changes. The requirements for military SAR application are very high resolution images of relatively small areas. The image must be provided with a very short time for any area of the world.
In this work aspects of high resolution spaceborne synthetic aperture radar with beam steering have been presented. Through the beam steering the performance of the radar is enhanced. The synthetic aperture radar system consists of two important components: the hardware and the processing algorithm. The system hardware including the transmitter, the antenna system and the receiver is described in the first part of this work. The processing algorithm is described in the second part of this work. The SAR raw receive signal resembles noise signal. It must be processed accordingly before it can be evaluated. Both, the hardware and the processing algorithm, influence the performance of the resulting radar image. Two SAR systems are considered in this work. The High-Resolution Wide-Swath (HRWS) SAR uses the Digital Beam Forming (DBF) to resolve the contradiction between the large azimuth coverage and the high azimuth resolution. The second considered system is a SAR working in the sliding spotlight mode. The ScanSAR mode is optimized for wide coverage by the price of reduced azimuth resolution. The spotlight mode provides radar images with high azimuth resolution but limited in swath extension. In the sliding spotlight mode the antenna beam is steered during the data acquisition to achieve a compromise between the high azimuth resolution and the wide coverage. Both considered systems use beam steering in order to enhance the resulting radar image. In this work presented are aspects of hardware and processing algorithm of a SAR system.
The first part of this work concerns aspects of the hardware design. Considered is the HRWS SAR system with multiple receive channels. In this system

the DBF in range and in azimuth is used to improve the performance of the system and to resolve the contradiction between the azimuth resolution and the coverage area. The full DBF provides high agility and improves significantly the performance of the radar system. The receive antenna is designed as an M by N antenna array. Each of the receive sub-apertures is followed by its own receive channel. To describe receiver front-end a new representation is proposed. The VKA-representation in frequency domain can describe in an uncomplicated and efficient way all components of the receive front-end, including the nonidealities and coupling between I and Q paths. The possibility to describe both I and Q paths with the same tree parameters of the VKA-representation is of advantage for simulation design and it's implementation. All components and effects can be described in the same way. In Chapter 3 the representation was used to model possible nonidealities of a multi-channel SAR receiver and its influence on the performance in a HRWS SAR simulation.

Minor position errors can influence the performance of a SAR system. This effect is considered in Chapter 4. The symmetrical and unsymmetrical bend of the antenna array plane have been investigated on an example of the HRWS SAR. In this system a narrow receive beam is created by the scan-on-receive DBF algorithm and slided over the ground. The symmetrical bend has more influence on the antenna gain of the array. The unsymmetrical bend causes less gain loss but the main lobe is shifted. Both types of antenna defects degrade the performance of the HRWS SAR system. The radar beam is not optimally adjusted for the direction of incoming reflected signal. The influence is stronger for narrower beam.

The high number of independent receivers results in high amount of raw data to be transmitted to the ground station or/and be processed. A possibility to reduce the amount of data is to use fewer bits for quantization. The original signal is then reconstructed with the DBF algorithm. This idea is investigated in Chapter 5. A special interest was put on the single bit quantization, were the ADC can be replaced by a simple comparator. The principle of one bit quantization and reconstruction by DBF is presented on an example of the HRWS SAR demonstrator. It was shown that additional white noise is necessary for the correct reconstruction on the receive signal in the low bit quantization. The multi-channel SAR system was simulated with various parameter configurations. The influence of the number of receive apertures, the input signal-to-noise ratio and the number of quantization bits had been investigated. The one bit quantization introduces a certain loss of performance which cannot be compensated by a higher number of receive sub-apertures. The loss of about 2 dB in signal-to-noise of the resulting radar image is not acceptable in case of a spaceborne SAR system. The optimal number of quantization bits for a multi-channel SAR system according to simulation results is four. This number of quantization ensures reconstruction of the radar signal with DBF without performance loss.

The second part of the work considers the processing algorithm of a synthetic aperture radar. The sliding spotlight mode is considered. This mode provides a compromise between the azimuth resolution and the length of the swath in azimuth. Longer Doppler history is achieved with an adequate beam steering in azimuth. The airborne case of the sliding spotlight mode is presented in Chapter 6. Two processing algorithms for sliding spotlight mode SAR based on the chirp scaling and the omega-K algorithm are proposed. The performance of both algorithms had been evaluated by the means of simulation. Both processing algorithms show good performance for the sliding spotlight mode data simulation up to 0.3m resolution. A proposal for algorithm correction for the antenna beam pattern in sliding spotlight mode is described. This algorithm corrects influence of the antenna pattern sliding spotlight mode data. The chirp scaling sliding spotlight algorithm is suggested for raw data with resolutions in range and azimuth of 0.5 m resolution. The omega-K sliding spotlight algorithm provides a slightly better performance for resolution of 0.3 m.

The spaceborne sliding spotlight mode is described in Chapter 7. Both proposed algorithms based on the chirp scaling and the omega-K algorithm, are modified to cope with spaceborne sliding spotlight mode raw data. The performance of both algorithms is evaluated in simulation with ideal round orbit. The processor based on the chirp scaling algorithm works satisfactory for a resolution of 0.5 m and 0.3 m. In the second simulation some slight degradation of processor performance is visible. The processor based on the omega-K algorithm is evaluated with the same data set. The performance of the processor is very good for both 0.5 m and 0.3 m resolution data. An additional simulation with 0.2 m has been conducted for the omega-K based processor. The performance for this high resolution is no longer acceptable for the proposed algorithm. The spaceborne sliding spotlight mode SAR processor based on omega-K algorithm provided better performance for highly resolved radar data. The processor based on chirp scaling algorithm provided

good performance with less necessary computation power for a resolution of 0.5 m. Both algorithms were modified to cope with the real orbit of a satellite. The proposed algorithm assumes that the position of the satellite is known or can be calculated for all acquisition points. The modified processor based on the omega-K algorithm is evaluated with simulations. The algorithm shows very good performance.
This work concerns aspects of the hardware and processing algorithms of a spaceborne synthetic aperture radar. It was shown that both parts are important to ensure best performance of the SAR system.

Bibliography

[1] *Microwaves and Radar Institute, Status Report 2000-2005, Research Results and Projects.* German Aerospace Center, April 2006.

[2] "Macdonald, Dettwiler and associates Ltd. (MDA)," Juni 2012. [Online]. Available: http://gs.mdacorporation.com/Default.aspx

[3] Astrium 2012, "TerraSAR–X: Astrium geoinformation services," Juni 2012. [Online]. Available: http://www.astrium-geo.com/terrasar-x/

[4] D. P. Belcher and C. J. Baker, "High resolution processing of hybrid strip-map spotlight mode SAR," *IEEE Proceeding on Radar and Sonar Navigation*, vol. 143, pp. 366–374, 1996.

[5] U. Benz, K. Strodl, and A. Moreira, "A comparison of several algorithms for SAR raw data compression," *IEEE Transactions on Geoscience and Remote Sensing*, vol. 33, pp. 1266–1276, 1995.

[6] S. Buckreuss, R. Werninghaus, and W. Pitz, "The german satellite mission TerraSAR-X," *IEEE Aerospace and Electronic Systems Magazine*, vol. 24, pp. 4–9, 2009.

[7] Bundesamt für Wehrtechnik und Beschaffung, "Internet page of Bundesamt für Wehrtechnik und Beschaffung," May 2010. [Online]. Available: http://www.bwb.org

[8] W. G. Carrara, R. S. Goodman, and R. M. Majewski, *Spotlight Synthetic Aperture Radar: Signal Processing Algorithms.* Artech House, Inc., 1995.

[9] J. Chen, Z. Wen, and Y. Zhou, "Analysis of ambiguity for synthetic aperture radar with mechanical distortion in phased array antenna," *International Geosience and Remote Sensing Symposium, IGARSS'05*, vol. 6, pp. 4049–4052, 2005.

[10] D. J. Cichon and W. Wiesbeck, "The heinrich hertz wireless experiments at karlsruhe in the view of modern communication," *International Conference on 100 Years of Radio*, pp. 1–6, 1995.

[11] R. Colburn, "Sound the alarm: A history of disaster detection and warning technologies," *IEEE The institute*, Sept. 2013.

[12] I. G. Cumming and F. H. Wong, *Digital Processing of Synthetic Aperture Radar Signal Data.* Artech House, Inc., 2005.

[13] A. Currie, "Synthetic Aperture Radar," *Electronics and Communication Engineering Journal*, vol. 3, pp. 159–170, 1991.

[14] A. Currie and M. A. Brown, "Wide-swath SAR," *IEEE Proceedings on Radar and Signal Processing*, vol. 139, pp. 122–135, 1992.

[15] N. G. S. F. C. Dr. David R. Williams, "Magellan Mission to Venus," February 2005. [Online]. Available: http://nssdc.gsfc.nasa.gov/planetary/magellan.html

[16] R. Duren, E. Wong, B. Breckenridge, S. Shaffer, C. Duncan, E. Tubbs, and P. Salomon, "Metrology, attitude, and orbit determination for spaceborne interferometric synthetic aperture radar," *Proceedings of SPIE, Acquisition, Tracking, and Pointing XII*, vol. 3364, pp. 51–60, 1998.

[17] Y. C. Eldar and A. V. Oppenheim, "Filterbank reconstruction of bandlimited signals from nonuniform and generalized samples," *IEEE Transactions on Signal Processing*, vol. 48, pp. 2864–2874, 2000.

[18] ESA Earthnet Online 2000-2012, "ESA Earth Online, ESA Future Mission GMES – Sentinels, Sentinel-1," Juni 2012. [Online]. Available: https://earth.esa.int/web/guest/missions/esa-future-missions/sentinel-1

[19] ESA Earthnet Online 2000-2012, "ESA Earth Online, Mission Cyrosat," Juni 2012. [Online]. Available: https://earth.esa.int/-web/guest/missions/esa-operational-eo-missions/cryosat

[20] ESA Earthnet Online 2000-2012, "ESA Earth Online, Mission Envisat," Juni 2012. [Online]. Available: https://earth.esa.int/-web/guest/missions/esa-operational-eo-missions/envisat

[21] J. P. Fitch, *Synthetic Aperture Radar*. Springer-Verlag New York Inc., 1988.

[22] G. Fornaro, R. Lanari, E. Sansosti, and M. Tesauro, "A two-step spotlight SAR data focusing approach," *International Geoscience and Remote Sensing Symposium, IGARSS'00*, vol. 1, pp. 84–86, 2000.

[23] G. Franceschetti, V. Pascazio, and G. Schirinzi, "Processing of signum coded sar signal: theory and experiments," *IEE Proceedings of Radar and Signal Processing*, vol. 138, pp. 192–198, 1991.

[24] N. Gebert, "Multi-channel azimuth processing for high-resolution wide-swath SAR imaging," Ph.D. dissertation, Forschungsberichte aus dem Institut für Höchstfrequenztechnik und Elektronik der Universität Karlsruhe, 2009.

[25] N. Gebert, G. Krieger, and A. Moreira, "Unambiguous SAR signal reconstruction from nonuniform displaced phase centre sampling," *International Geoscience and Remote Sensing Symposium, IGARSS'04*, vol. 1, pp. 260–264, 2004.

[26] N. Gebert, G. Krieger, and A. Moreira, "Multi-channel azimuth processing for high-resolution wide-swath SAR imaging - system performance and influence of perturbations," *International Radar Symposium, IRS'05*, pp. 49–54, 2005.

[27] N. Gebert, G. Krieger, and A. Moreira, "SAR signal reconstruction from non-uniform displaced phase centre sampling in the presence of perturbations," *International Geoscience and Remote Sensing Symposium, IGARSS'05*, vol. 2, pp. 1034–1037, 2005.

[28] N. Gebert, G. Krieger, and A. Moreira, "Digital beamforming for HRWS-SAR imaging: System design, performance and optimization strategies," *International Geoscience and Remote Sensing Symposium, IGARSS'06*, pp. 1836–1839, 2006.

[29] N. Gebert, G. Krieger, and A. Moreira, "Digital beamforming on receive: Techniques and optimization strategies for High-Resolution Wide-Swath SAR imaging," *IEEE Transactions on Aerospace and Electronic Systems*, vol. 45, pp. 564–592, 2009.

[30] German Aerospace Center (DLR), "Space operations and astronaut training: Space missions: Current missions: Sar-lupe," February 2013. [Online]. Available: http://www.dlr.de/rb/en/desktopdefault.aspx

[31] C. Heer and P. F. Shutie, "Digital beam forming synthetic aperture radar," *European Microwave Conference, EuMC'04*, vol. 3, pp. 1513–1516, 2004.

[32] C. Heer, F. Soulalle, R. Zahn, and R. Reber, "Investigations on a new high resolution wide swath SAR concept," *International Geoscience and Remote Sensing Symposium, IGARSS'03*, vol. 1, pp. 521–523, 2003.

[33] C. Heer, C. Fischer, and C. Schaefer, "Spaceborne SAR systems and technologies," *International Microwave Symposium Digest (MTT)*, pp. 538–541, 2010.

[34] C. R. Jackson and J. R. Apel, *Synthetic Aperture Radar Marine User's Manual.* U.S. Department of Commerce, 2004.

[35] Jet Propulsion Laboratory California Institute of Technology, "Magellan Mission to Venus (JPL)," Juni 2012. [Online]. Available: http://www2.jpl.nasa.gov/magellan/

[36] Jet Propulsion Laboratory California Institute of Technology, "What is SIR-C/X-SAR?" February 2013. [Online]. Available: http://southport.jpl.nasa.gov/desc/SIRCdesc.html

[37] T. Kempf, H. Anglberger, and H. Süß, "Depth-of-focus issues on spaceborne very high resolution SAR," *International Geoscience and Remote Sensing Symposium, IGARSS'12*, pp. 7448–7451, 2012.

[38] J. Kim, D. Becker, W. Wiesbeck, and M. Younis, "Experimental performance analysis of digital beam forming on synthetic aperture radar," *European Conference on Synthetic Aperture Radar, EUSAR'08*, pp. 1–4, 2007.

[39] J. Kim, A. Ossowska, and W. Wiesbeck, "Experimental investigation of digital beamforming sar performance using a ground-based demonstrator," *International Geoscience and Remote Sensing Symposium, IGARSS'07*, pp. 111–114, 2007.

[40] J. Kim, M. Younis, and W. Wiesbeck, "Experimental performance investigation of digital beamforming on synthetic aperture radar," *International Geoscience and Remote Sensing Symposium, IGARSS'08*, vol. 1, pp. 176–176, 2007.

[41] H. Klausing and W. Holpp, *Radar mit realer und synthetischer Apertur, Konzeption und Realisierung.* Oldebbourg, 1999.

[42] H. J. Kramer, "Sharing Earth Observation Resources," January 2008. [Online]. Available: http://www.eoportal.org/directory/-pres_SRTMShuttleRadarTopographyMission.htm

[43] G. Krieger, N. Gebert, and A. Moreira, "SAR signal reconstruction from non-uniform displaced phase centre sampling," *International Geoscience and Remote Sensing Symposium, IGARSS'04*, vol. 3, pp. 1763–1766, 2004.

[44] G. Krieger, N. Gebert, M. Younis, F. Bordoni, A. Patyuchenko, and A. Moreira, "Advanced concepts for ultra-wide-swath SAR imaging," *European Conference on Synthetic Aperture Radar, EUSAR'08*, pp. 1–4, 2008.

[45] G. Krieger, N. Gebert, M. Younis, F. Bordoni, A. Patyuchenko, and A. Moreira, "Multidimensional waveform encoding: A new digital beamforming technique for synthetic aperture radar remote sensing," *IEEE Transactions on Geoscience and Remote Sensing*, vol. 46, pp. 31–46, 2008.

[46] G. Krieger and A. Moreira, "Potentials of digital beamforming in bi- and multistatic SAR," *IEEE Proceedings on Radar, Sonar and Navigation*, vol. 153, pp. 184–198, 2006.

[47] G. Krieger, M. Younis, N. Gebert, S. Huber, F. Bordoni, A. Patyuchenko, and A. Moreira, "Advanced concepts for high-resolution wide-swath SAR imaging," *European Conference on Synthetic Aperture Radar, EUSAR'10*, pp. 1–4, 2010.

[48] C. Kuhnert, "Systemanalyse von Mehrantennen Frontends (MIMO)," Ph.D. dissertation, Forschungsberichte aus dem Institut für Höchstfrequenztechnik und Elektronik der Universität Karlsruhe, 2006.

[49] R. Lanari, S. Zoffoli, E. Sansosti, G. Fornaro, and F. Serafino, "New approach for hybrid strip-map/spotlight SAR data focusing," *IEEE Proceeding on Radar and Sonar Navigation*, vol. 148, pp. 1350–2395, 2001.

[50] R. Lanari, M. Tesauro, E. Sansosti, and G. Fornaro, "Spotlight SAR data focusing based on a two-step processing approach," *IEEE Transactions on Geoscience and Remote Sensing*, vol. 39, pp. 1993–2004, 2001.

[51] R. G. Lyons, *Wprowadzenie do cyfrowego przetwarzania sygnałów*. Wydawnictwa Komunikacji i Łączności sp. z o.o., 2000.

[52] A. Meta, P. Hoogeboom, and L. P. Ligthart, "Sampling quantization analysis and results for FMCW SAR," *International Geoscience and Remote Sensing Symposium, IGARSS'07*, pp. 1029–1032, 2007.

[53] J. Mittermayer, "Hochauflösende Verarbeitung von Radardaten mit Synthetischen Apertur," Ph.D. dissertation, Department of Electrical Engineering and Computer Science, 2000.

[54] J. Mittermayer, R. Lord, and E. Borner, "Sliding spotlight SAR processing for TerraSAR-X using a new formulation of the extended chirp scaling algorithm," *International Geoscience and Remote Sensing Symposium, IGARSS 2003*, vol. 3, pp. 1462–1464, 2003.

[55] A. Moreira, G. Krieger, I. Hajnsek, D. Hounam, M. Werner, S. Riegger, and E. Settelmeyer, "TanDEM-X: A TerraSAR-X add-on satellite for single-pass SAR interferometry," *International Geoscience and Remote Sensing Symposium, IGARSS'04*, vol. 2, pp. 1000–1003, 2004.

[56] S. Ochs and W. Pitz, "The TerraSAR-X and TanDEM-X satellites," *International Conference on Recent Advances in Space Technologies, RAST'07. 3rd*, pp. 294–298, 2007.

[57] E. Oleary, A. Richardson, B. Chapman, and A. Wong, "SEASAT 1978," February 1998. [Online]. Available: http://southport.jpl.nasa.gov/scienceapps/seasat.html

[58] A. V. Oppenheim, R. W. Schafer, and J. R. Buck, *Discrete-time signal processing (2nd ed.)*. Upper Saddle River, NJ, USA: Prentice-Hall, Inc., 1999.

[59] A. Ossowska, "Auswahl von Objekten aus SAR-Bildern mit optimaler Übereinstimmung zu idealen Punktzielen zur Nutzung f einen Phasen-Gradient Autofocus (PGA) Algorithmus," *Diplomarbeit, Institut für Höchstfrequenztechnik und Elektronik, Universität Karlsruhe (TH)*, 2004.

[60] A. Ossowska, J. H. Kim, and W. Wiesbeck, "Influence of antenna array distortion on the performance of the HRWS SAR," *Advanced RF Sensors for the Earth Observation, ESTEC'06*, 2006.

[61] A. Ossowska, J. H. Kim, and W. Wiesbeck, "A simulator for synthetic aperture radar with digital beam forming processing in elevation," *International Geoscience and Remote Sensing Symposium, IGARSS 2006*, pp. 1407–1410, 2006.

[62] A. Ossowska, J. H. Kim, and W. Wiesbeck, "Influence of mechanical antenna distortions on the performance of the HRWS SAR system," *International Geoscience and Remote Sensing Symposium, IGARSS 2007*, pp. 2152–2155, 2007.

[63] A. Ossowska, J. H. Kim, and W. Wiesbeck, "Modeling of nonidealities in receiver front-end for a simulation of multistatic SAR system," *European Radar Conference, EuRAD'07*, pp. 13–16, 2007.

[64] A. Ossowska, J. H. Kim, and W. Wiesbeck, "Simulation of high-resolution wide-swath SAR," *European Conference on Synthetic Aperture Radar, EuSAR'06*, 2006.

[65] A. Ossowska and R. Speck, "Hybrid strip-map/spotlight mode processing based on chirp scaling processing," *International Radar Symposium, IRS'08*, 2008.

[66] A. Ossowska and R. Speck, "Processing of sliding spotlight mode data with consideration of orbit geometry," *Signal Processing Symposium, SPS'09*, 2009.

[67] A. Papoulis, "Generalized sampling expansion," *IEEE Transaction on Circuits and Systems*, vol. 24, 1977.

[68] V. Pascazio and G. Schirinzi, "Signum coded synthetic aperture radar: The effect of oversampling on image quality," *International Conference on Image Processing*, vol. 1, pp. 849–852, 1997.

[69] V. Pascazio and G. Schirinzi, "Synthetic aperture radar imaging by one bit coded signals," *Electronics & Communication Engineering Journal*, vol. 10, pp. 17–28, 1998.

[70] P. Prats, R. Scheiber, J. Mittermayer, A. Meta, and A. Moreira, "Processing of sliding spotlight and tops sar data using baseband azimuth scaling," *IEEE Transactions on Geoscience and Remote Sensing*, vol. 48, pp. 770–780, 2009.

[71] P. Prats, R. Scheiber, J. Mittermayer, and A. Moreira, "Processing multiple SAR modes with baseband azimuth scaling," *International Geoscience and Remote Sensing Symposium, IGARSS 2009*, vol. 5, pp. 172–175, 2009.

[72] A. Prokes, "Generalized sampling theorem for bandpass signals," *EURASIP Journal on Applied Signal Processing*, 2006.

[73] L. R. Rabiner and B. Gold, *Theory and application of digital signal processing*. Prentice-Hall, Inc. Englewood Cliffs, New Jersey, 1975.

[74] E. Ramirez, "Jet Propulsion Laboratory, California Institute of Technology," August 2005. [Online]. Available: http://www2.jpl.nasa.gov/srtm/instr.htm

[75] R. Raney, "Considerations for SAR image quantification unique to orbital system," *IEEE Transaction on Geoscience and Remote Sensing*, vol. 29, pp. 754–760, 1991.

[76] R. K. Raney, H. Runge, R. Bamler, I. Cumming, and F. H. Wong, "Precision SAR processing using chirp scaling," *IEEE Transaction on Geoscience and Remote Sensing*, vol. 32, pp. 786–799, 1994.

[77] M. A. Richards, "A beginner's guide to interferometric SAR concepts and signal processing [AESS tutorial IV]," *IEEE Aerospace and Electronic Systems Magazine*, vol. 22, pp. 5–29, 2006.

[78] B. Schiek, *Grundlagen der Hochfrequenz-Messtechnik.* Springer-Verlag Berlin Heildelberg New York, 1999.

[79] D. Schrank, "Optimale Quantisierung stark verrauschter Radarsignale," *Studienarbeit, Institut für Höchstfrequenztechnik und Elektronik, Universität Karlsruhe (TH)*, 2006.

[80] J. Schulz-Stellenfleth, S. Lehner, T. K. nig, A. Reppucci, and S. Brusch, "Use of tandem pairs of ERS-2 and ENVISAT SAR data for the analysis of oceanographic and atmospheric processes," *International Geoscience and Remote Sensing Symposium, IGARSS'07*, pp. 3265–3268, 2007.

[81] R. Speck, B. Devereux, G.-P. Schumann, J. Neal, and P. Bates, "Flood detection in urban areas using TerraSAR-X," *IEEE Transactions on Geoscience and Remote Sensing*, vol. 48, pp. 882–894, 2010.

[82] C. Sturm, "Gemeinsame Realisierung von Radar-Sensorik und Funkkommunikation mit OFDM-Signalen," Ph.D. dissertation, Forschungsberichte aus dem Institut für Höchstfrequenztechnik und Elektronik der Universität Karlsruhe, 2011.

[83] M. Suess, B. Grafmueller, and R. Zahn, "A novel high resolution, wide swath SAR system," *International Geoscience and Remote Sensing Symposium, IGARSS'01*, vol. 3, pp. 1914–1917, 2001.

[84] M. Suess, M. Zubler, and R. Zahn, "Performance investigation on the high resolution, wide swath SAR system," *European Conference on Synthetic Aperture Radar, EuSAR'02*, pp. 187–190, 2002.

[85] M. Suess, M. Zubler, and R. Zahn, "Performance investigation on the high resolution, wide swath SAR system," *International Geoscience and Remote Sensing Symposium, IGARSS'03*, vol. 153, pp. 184–198, 2006.

[86] H. Süß, R. Schröder, M. Peichl, and T. Neff, "Possible military requirements and applications of active and passive imaging sensors at micro- and millimeterwave frequencies," *International Geoscience and Remote Sensing Symposium, IGARSS'02*, 2002.

[87] S. S. Swords, *Technical history of beginnings of RADAR*. Peter Peregrinus Ltd, 1986.

[88] J. Szóstka, *Fale i anteny*. Wydawnictwa Komunikacji i Łączności sp. z o.o., 2001.

[89] A. R. Thompson, J. M. Moran, and G. W. Swenson, Jr., *Interferometry and Synthesis in Radio Astronomy*. John Wiley & Sons, Inc., 1986.

[90] M. Thumm, W. Wiesbeck, and S. Kern, *Hochfrequenzmesstechnik Verfahren und Messsysteme*. Vieweg+Teubner | GWV Fachverlage GmbH, 1998.

[91] U. Tietze and C. Schenk, *Układy półprzewodnikowe*. Wydawnictwa Naukowo-Techniczne, 1996.

[92] B. van den Broek, E. den Breejen, R. Dekker, and A. Smith, "Change detection and maritime situation awareness in the channel area - feasibility of space borne SAR for maritime situation awareness," *International Geoscience and Remote Sensing Symposium, IGARSS'12*, pp. 7436–7439, 2012.

[93] H. Wang, "Performance of phased-array antennas with mechanical errors," *IEEE Transactions on Aerospace and Electronic Systems*, vol. 28, pp. 535–545, 1992.

[94] Y. Wang and D. Yao, "Analysis of some key parameters in sliding spotlight SAR," *International Radar Conference, IET'09*, pp. 1–4, 2009.

[95] B. Widrow, I. Kollár, and M.-C. Liu, "Statistical theory of quantization," *IEEE Transactions on Instrumentation and Measurement*, vol. 45, pp. 353–361, 1996.

[96] W. Wiesbeck, *Lecture Script: Radar System Engineering (Edition 16th)*. Universität Karlsruhe, Institut für Hochfrequenztechnik und Elektronik, 2009.

[97] M. Younis, "Digital beam-forming for high resolution wide swath real and synthetic aperture radar," Ph.D. dissertation, Forschungsberichte aus dem Institut für Höchstfrequenztechnik und Elektronik der Universität Karlsruhe, 2004.

[98] M. Younis, J. Boer, C. Ortega, D. Schulze, S. Huber, and J. Mittermayer, "Determening the optimum compromise between SAR data compression and radiometric performance -an approach based on the analysis of TerraSAR-X data-," *International Geoscience and Remote Sensing Symposium, IGARSS'08*, vol. 3, pp. 107 – 110, 2008.

[99] M. Younis, C. Fischer, and W. Wiesbeck, "An evaluation of performance parameters of reconfigurable SAR systems," *International Geoscience and Remote Sensing Symposium, IGARSS'02*, vol. 1, pp. 677–679, 2002.

[100] M. Younis, C. Fischer, and W. Wiesbeck, "Digital beamforming in SAR systems," *IEEE Transaction on Geoscience and Remote Sensing*, vol. 41, pp. 1735–1739, 2003.

[101] M. Younis, G. Krieger, and A. Moreira, "MIMO SAR techniques and trades," *European Radar Conference, EuRAD'13*, pp. 141–144, 2013.

[102] M. Younis and W. Wiesbeck, "SAR with digital beamforming on reciever only," *International Geoscience and Remote Sensing Symposium, IGARS' 99*, vol. 41, pp. 1773–1775, June 1999.

[103] M. Younis and W. Wiesbeck, "Antenna system for a forward looking SAR using digital beamforming on-receive-only," *International Geoscience and Remote Sensing Symposium, IGARSS'00*, vol. 5, pp. 2343–2345, 2000.

[104] V. Zamparelli, G. Fornaro, R. Lanari, S. Perna, and D.Reale, "Processing of sliding spotlight SAR data in presence of squint," *International Geoscience and Remote Sensing Symposium, IGARSS'12*, pp. 2137–2140, 2012.

[105] A. Zielska, "Entwicklung eines Omega-K Algorithmus zur Processierung von SAR-Rohdaten für den Strip-Map/Spotlight Hybrid-Modus," *Diplomarbeit, Institut für Höchstfrequenztechnik und Elektronik, Universität Karlsruhe (TH)*, 2008.